ACS SYMPOSIUM SERIES 782

Gas Chromatography–Olfactometry

The State of the Art

Jane V. Leland, EDITOR
Kraft Foods

Peter Schieberle, EDITOR
Deutsche Forchungsanstalt für Lebensmittelchemie

Andrea Buettner, EDITOR
Deutsche Forchungsanstalt für Lebensmittelchemie

Terry E. Acree, EDITOR
Cornell University

American Chemical Society, Washington, DC

Library of Congress Cataloging-in-Publication Data

Gas chromatography–olfactometry : the state of the art / Jane V. Leland...[et al.], editors,

   p. cm.—(ACS symposium series ; 782)

   ISBN 0–8412–3713–1

   1. Gas chromatography—Congresses. 2. Olfactometry —Congresses.

   I. Leland, Jane V. II. Series.

QD79.C45 G366   2001
543′.0896—dc21   00–50414

The paper used in this publication meets the minimum requirements of American National Standard for Information Sciences—Permanence of Paper for Printed Library Materials, ANSI Z39.48–1984.

Foreword

The ACS Symposium Series was first published in 1974 to provide a mechanism for publishing symposia quickly in book form. The purpose of the series is to publish timely, comprehensive books developed from ACS sponsored symposia based on current scientific research. Occasionally, books are developed from symposia sponsored by other organizations when the topic is of keen interest to the chemistry audience.

Before agreeing to publish a book, the proposed table of contents is reviewed for appropriate and comprehensive coverage and for interest to the audience. Some papers may be excluded to better focus the book; others may be added to provide comprehensiveness. When appropriate, overview or introductory chapters are added. Drafts of chapters are peer-reviewed prior to final acceptance or rejection, and manuscripts are prepared in camera-ready format.

As a rule, only original research papers and original review papers are included in the volumes. Verbatim reproductions of previously published papers are not accepted.

ACS Books Department

Dedication

We are extremely pleased to dedicate this book to Professor Dr. Werner Grosch in recognition of his longstanding service and his enduring achievements to the field of flavor research. Professor Grosch embarked on his groundbreaking career in the field with his first publications on the analysis of volatile compounds in tomatoes in the early 1960s. As of 1999, Professor Grosch is Professor emeritus with the Deutsche Forschungsanstalt für Lebensmittelchemie in Garching, Germany.

JANE V. LELAND
Kraft Foods, Inc.
801 Waukegan Road
Glenview, IL 60025

PETER SCHIEBERLE
Deutsche Forschungsanstalt
  für Lebensmittelchemie
Lichtenbergstrasse 4
D–86748 Garching
Germany

ANDREA BUETTNER
Deutsche Forschungsanstalt
  für Lebensmittelchemie
Lichtenbergstrasse 4
D–86748 Garching
Germany

TERRY E. ACREE
Department of Food Science
  and Technology
Cornell University
New York State Agricultural
  Experiment Station
Geneva, NY 14456–044462

Contents

Preface

Smell figures prominently in our enjoyment of life. Flavors and aromas contribute much to the appeal of food, whereas perfumes or fragrances can evoke powerful emotions and memories. Unquestionably then, flavors and fragrances are important to consumers. As a consequence, they are also important to researchers, developers, and manufacturers of foods, perfumes, and household and beauty products.

The Holy Grail remains elusive: the discovery of which volatile compounds really matter to our overall perception of the complex mixtures that typify natural flavors and fragrances. Identification and quantification of such aroma- or odor-active compounds has many important motivations, including the following:

- correlating sensory responses with volatile chemicals

- selecting compounds useful for monitoring flavor changes during shelf life, processing, and so on

- resolving off-flavor problems

- understanding flavor release during eating

- augmenting creative flavor compounding

- targeting flavor compounds for thermal or biological generation

- assessing olfactory acuity of individuals

- elucidating the mechanism of odor receptors

Gas Chromatography–Olfactometry or GC–O refers to the sniffing of GC effluent to determine which components possess odor. Many of the peaks detected by the GC do not actually contribute to our perception of flavors or fragrances, because they are present below our thresholds for detecting them. The emergence of GC–O in the 1960s and 1970s was a landmark development in flavor, aroma, and fragrance research. It provides valuable information about the areas of a gas chromatogram on which to focus attention and resources. In

ix

absence of such guidance, attempting to identify all the peaks in a gas chromatogram is a monumental, if not futile, endeavor.

In 1963 Rothe introduced the concept of Odor Activity Value (OAV) as the concentration of an odorant relative to its human threshold. OAVs are thus an indication of the potency of a specific odorant in a specific sample and have proven to be extremely useful. The next pivotal advance came in the mid 1980s when the Acree Laboratory of Cornell University and the Grosch laboratory of the Technical University of Munich independently developed similar GC–O-based techniques for estimating OAVs: CharmAnalysis and Aroma Extract Dilution Analysis (AEDA). These similar, albeit differing, techniques have become the mainstays for characterizing key odorants in modern aroma research.

This book is the first compilation of research on the subject of GC–O. It is based on the symposium, "Advances in Gas Chromatography–Olfactometry", held at the 219th American Chemical Society (ACS) National Meeting in New Orleans, Louisiana during August of 1999. The impetus for the symposium was to provide information and stimulate interchange about the latest applications and developments in the field of GC–O. The contributors to the symposium and this volume include a select group of internationally renowned experts in this highly specialized field and offer the most comprehensive and current perspective in GC–O. As such, the book will be of special interest as a state-of-the-art reference for industrial and academic flavor and fragrance researchers as well as students of flavor, fragrance, and food, but it can also serve as an introduction for new comers to the field of GC–O.

The book focuses on a broad array of applications of the GC–O techniques first developed in the mid 1980s in addition to highlighting the latest novel developments. Applications include the use of standard CharmAnalysis and AEDA to characterize key flavor or aroma compounds, as well as extensions that couple GC–O with modern sampling techniques such as Solid-Phase MicroExtraction (SPME) in an attempt to simplify the process. Chapters also address the application of GC–O to flavor creation, the use of cross-modality matching for estimating odor intensity, and the recent novel idea of using GC–O to screen individuals for differences in individual olfactory acuity.

Second generation techniques presented in the book include Aroma Extract Concentration Analysis (AECA) and sample dilution. The book also explores some limitations of the use of odor activity values, the implications of which should open our views on correlating perceived intensity with concentration for important flavor and aroma volatiles.

With such powerful tools at their disposal, today's flavor chemists are able to forge ahead productively in overturning new, important, and exciting flavor and fragrance compounds. However, it is not enough to simply identify these key compounds—the next challenge is to find better ways to establish links between key compounds and sensory–consumer perceptions.

Acknowledgments

I extend special appreciation to Professor Terry Acree for co-organizing the original symposium with me, and to all the scientists who made the symposium such a success by contributing stimulating presentations and discussion. Kraft Foods is acknowledged for its generous financial sponsorship of the symposium. Thanks are also due to the contributing authors for their time and effort in preparing chapters for this book, as well as those scientists who served as referees for the chapters. I feel extremely fortunate to have had such conscientious coeditors with whom it was a pleasure to work. Last, but not least, I thank my husband Rich Fisher for his support and flexibility.

JANE V. LELAND
Kraft Foods, Inc.
801 Waukegan Road
Glenview, IL 60025

Chapter 1

Advances in Gas Chromatography–Olfactometry

A. R. Mayol[1] and T. E. Acree[2,3]

[1]Consorzio Ricerca Filiera Lattiero-Casearia, Ragusa, Italia
[2]New York State Agricultural Experiment Station, Food Science and
Technology, Cornell University, Geneva, NY 14456
[3]Corresponding author

This paper compiles an overview of the development of gas
chromatography-olfactometry (GC-O), basic GC-O principles and
the future of this technique and its potential applications in industry
and academia. Early aroma isolation studies of food characterized
food extracts by gas chromatography-mass spectrometry (GC-MS)
and sensory evaluations. With the development of GC-O, the
identification of odor active compounds and quantification of odor
activity was achieved while the extraction techniques used a priori
were improved. The Flavornet, a database composed of over 500
odor active compounds compiled from GC-O data, will be useful in
the understanding of the odor receptors' mechanism and the
correlation of sensory experiments with chemical responses.

When humans experience the flavor of a mixture of chemicals they are not able
to distinguish most of the component flavorants that make up the experience. Their
response is an integration of the pattern of stimulants into a pattern of perceptions
without a one-to-one correspondence between stimulant and distinguishable percept.
Descriptive sensory experiments will produce a multivariate description of these
perceptions that can be expressed in a spider diagram like the summary of Concord
grape juice flavor shown in Figure 1. Some of the components in Fig. 1 are olfactory
perceptions caused by volatiles while others are taste or chemesthetic responses to
non-volatile components. "Concord grape " flavor is a label for a temporal experience
of the several percepts shown in Fig. 1 that includes the time dependent suppression
of some of the components as well as the total inhibition of percepts not shown in the
figure. A flavor experience is a very short sensory "movie" summarized by a title.

The cast of characters are the percepts that are experienced and the actors are the stimulants that create the characters. Unfortunately, each character is played by several actors while some actors are never visible on stage at all.

Nevertheless, like actors in a movie all the flavor stimulants in food are detectable when they are experienced alone, especially at the levels at which they are present in food. For this reason gas chromatography / olfactometry (GC/O) can detect all the odorants that contribute to flavor, if they are properly sampled and chromatographed.

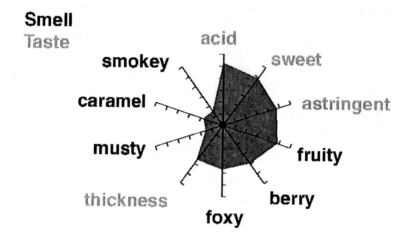

Figure 1. A spider diagram of the average sensory percepts for Concord Grape Juice (1).

Because spider diagrams are averaged over time and thus do not reveal the entire sensory experience, time-intensity studies must be combined with dynamic flavor release analyses for the complete story to be told. Too time consuming and complicated to use in most QC or research experiments, the temporal component is usually averaged out in descriptive analysis to produce a time average or temporal slice of the total experience. Although sensory experiments do address the psychometric description of the flavor experience, chemical analysis is needed to provide information about composition: the feature of food modifiable by processors. For example, the titratable acidity, pH, refractive index and specific gravity of a wine are used to predict and manipulate taste but the measurement and prediction of smell is not as well developed. GC/O, however, is providing indicators of aroma that can be monitored by chemical analysis, e.g. methyl anthranilate and o-aminoacetophenone in Labruscana, cis-rose oxide in Gewürztraminer grapes, etc. Figure 2 shows the typical result from a GC/O study of a natural product, peppermint oil. The most abundant compound shown in the FID at the bottom of Fig. 2 is menthol, showing very little odor potency in the GC/O chromatogram above. In contrast the most odor potent compound in the GC/O chromatogram shows very little response in the GC/FID. It is a fact that most odorants found in natural products, including both food ingredients and fragrance materials, are present at trace levels and are difficult to detect in the presence of more abundant and less potent components. GC/O has over the years provided the best means to focus isolation and identifications on the most potent odorants in natural products.

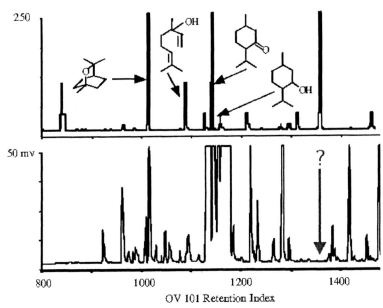

Figure 2. The GC/O (upper) and GC/FID (lower) of peppermint oil showing the high potency of a component present in only trace amounts in the extract.

History of GC/O

Soon after gas chromatography (GC) was invented (2), chemists interested in studying smell began to sniff the GC effluent to determine which components had odor. In 1964 Fuller, Steltenkamp, and Tisserand published the first description of a GC modified for this purpose (3). The design consisted of a non-destructive but insensitive thermal conductivity detector with the outlet sent to a sniff port located in a telephone booth to isolate the "sniffer" from other ambient smells. In 1971 the first "true" GC-olfactometer that mixed humid air with the GC effluent was reported (4). This design diminished the discomfort and inaccuracy associated with sniffing a stream of hot GC effluent. Further improvements included the use of a venturi to retain the resolution of capillary columns (5) and to deliver the effluent ergonomically to the "sniffer". The development of the GC-O as a quantitative bioassay (6-7) led to the recognition that fewer than a thousand compounds are in the pallet of odors that make up the olfactory stimulants found in food and fragrance (8).

In 1963 Rothe derived odor activity values (OAV) in order to correlate concentration of the odorant with detection threshold values (9). OAV is defined as concentration / threshold where the threshold is the lowest concentration detectable by humans in the sample. OAV indicates the potency of a specific odorant in a specific sample. Humans should not be able to detect chemicals that are at OAV levels lower that one. OAV values greater than one "may" be detectable depending on flavor release and odor suppression for other components in the sample. In general OAV values are the maximum an odorant can contribute to a mixture. However, OAV values do not correlate linearly with perceived intensity values and do not predict the odor intensity of compounds in combination. Moreover, threshold values in a particular system are seldom available.

Dilution factors and Charm values are two similar and effective means of estimating the OAV of compounds in completely vaporized samples. A dilution factor is the number of times, expressed as fold (7), that a sample can be diluted before an odor disappears from a GC/O chromatogram. Charm values (6), are measures of the areas under GC/O peaks which are derived from both the dilution and the duration of odor events. To account for flavor release, sampling techniques such as GCO-H (10), retronasal aroma simulation (RAS) (11), and isotopic standards in SIDA (12) have been used. SPME of isolated headspaces (13) and modified SPME lengths and thicknesses (14) have also simplified GC/O techniques. In order to use GC/O as a direct measure of relative odor intensity, methods based on the time-intensity assessment of GC effluents were developed (15) and applied to citrus flavor (16).

Recently, Stevens' law (17) was used to justify normalization and transformation of charm, flavor dilution factor(s)(FD), and OAV values for the purpose of comparison and to generate data that reflected the potential odor character of a mixture of odorants . Stevens' law is expressed as follows: $\Psi = k\ \Phi^n$ where Ψ is the perceived intensity of a stimulant, k is a constant, Φ is the stimulus level (FD factor(s), Charm, OAV), and n is a Stevens' exponent (0.3 – 0.8). Odor spectrum values (OSV) are potency values normalized to the most potent odorant in the sample and raised to the Stevens'exponent (0.3 – 0.8) or an average of 0.5. OSV's can be calculated from AEDA, Charm and OAV's from stable isotope dilution analysis (SIDA) data. This allows us to compare the data from entirely different kinds of experiments and predict the flavor experience in a similar way.

Sampling Techniques

Sampling technique remains one of the least developed aspects and major sources of bias in GC/O and GC/MS analysis of flavor. Odor-active compounds are released from food at different rates during eating, yielding patterns of composition that differ greatly from most compositions obtained from extraction and distillation techniques. While chemical manipulation tends to disrupt the ratio of odor-active chemicals present in the food systems, milder sampling techniques might yield extracts that are not concentrated enough. Moreover, some sampling techniques might incorporate thermally released or non-volatile artifacts to the extract. The behavior of the analytes in a sample matrix will dictate the most appropriate sampling technique since no single technique is appropriate for all samples. For example, simultaneous solvent extraction distillations (18) are typically time-consuming but they do not create certain artifacts. Solvent extraction is a faster more direct method in which artifact formation is caused by non-volatile compounds extracted from the sample and the loss of volatile components during concentration. When sequential extractions are performed using two solvents both polar and non-polar compounds can be analyzed. Headspace analysis is also a non-artifact forming method that is fast when it utilizes solid phase micro-extraction (SPME), thermal desorption, or direct injection. The most common bias caused by headspace analysis is the difference between dynamic and equilibrium headspace composition. Figure 3 shows a comparison between the dynamic and the static headspace above a cola beverage (14)

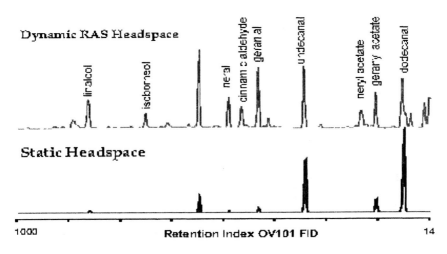

Figure 3 A comparison between dynamic and static headspace composition above a simulated cola beverage (14).

Static headspace approximates orthonasal aroma perceived directly through external nares when a sample is being sniffed. Retronasal aroma, the odor perceived when aroma compounds are released during mastication and salivation, is approximated by dynamic headspace. Retronasal aroma is transported to the lungs and then through the anterior nasal cavity to the olfactory epithilium. By simply holding your nose during eating you can see how important this route is to the enjoyment of food. Studies performed by Roberts, et. al. were geared towards the development of an improved purge and trap instrument called a retronasal aroma simulator (RAS). The RAS effectively controlled the four main factors affecting flavor release during eating: temperature, respiration, salivation and mastication (19,11).

Extending Roberts' work, Diebler found that odorants are released in different ratios when exposed to conditions similar to that of the mouth (14). Diebler studied the flavor release of a group of standards in solution where two independent measurements were taken. The static headspace of the artificial flavor was probed using solid phase micro-extraction (SPME). A dynamic measurement of the headspace was also undertaken utilizing the RAS developed by Roberts and Acree and SPME as the sampling device. Figure 3 clearly shows the difference between these two sampling methods.

Future of GC/O

Flavornet is an Internet-based database currently containing information on 550 odor active compounds that have been identified by GC/O (8). This database is useful for the identification of odor active components using retention indexes (RI) and odor authentic standards. Available at the following URL: http://www.nysaes.cornell.edu/fst/faculty/acree/flavornet , the data compiled includes Kovats RI's for OV101, DB5, OV1701, and Carbowax columns, ester RI's for each column, CAS number, common name, chemical name, general odor quality, specific odor quality, chemical structure, pdb three-dimensional structure files, and references. The Flavornet can be browsed by a chemical's RI's or odor quality and it is arranged according to both Kovats or ethyl esters index for each substrate. The RI's are color coded with the blue entries representing values as reported in the literature while the black RI's are values that, in the absence of experimental or reported data on the given substrate, have been interpolated according to formulas that simulate the retention properties for each compound on that substrate. The database is also arranged according to a lexicon containing 26 generic aroma categories based on ASTM DS-66 (20) and hundreds of specific aroma descriptors from the literature. At present the lexicon is based on published GC/O reports but the inclusion of "synonyms" and other languages will greatly improve its usefulness. To make this tool more powerful threshold values in air, water, and oil with the respective references are being developed. Another database-centered tool could be generated in the future, in which the odor profile (spectrum) of a particular food, essential oil, flower or reaction is compiled. An odor spectrum looks very much like a GC/MS mass spectrum plot but with the activity spikes representing the odor potency and retention index properties of each of the odor-active components in the matrix. These odor spectra could form the basis for an olfactory "finger print" database that would aid in the detection of off-flavors, the study of relatedness in biological material (e.g. by detecting olfactory markers or patterns common to related species of plants), and perhaps add valuable information to the study of the evolution of human olfactory function.

In the future GC-O data can be compared with databases like the Flavornet and verified with authentic standards to detect and *identify* off or in-appropriate odors without the need for GC/MS. Furthermore, the completion of the list of "natural" food and fragrance odorants may further diminish the need for more expensive spectroscopic methods. An opposing view however could hold that with most of the food and fragrance odorants known and both normal and isotopic standards available (still some way off) selected ion monitoring chemical ionization mass spectroscopy can become routine, reducing or even eliminating the need for relatively labor intensive methods such as GC/O in quality control and basic research laboratories.

Obtaining valid headspace data to correlate with compositional data is a most pressing challenge for flavor chemists. New headspace sampling techniques capable of efficiently undergoing dilution analysis in a reproducible fashion would go a long way toward helping us to understand flavor release during eating. The use of direct injection is limited due to the low concentration of odorants in the samples and the need for pre-concentration steps prior to injection. SPME is widely used since it is a clean, simple, solvent-free, and fast method for extracting volatiles from headspace. However applications of SPME to GCO are presently somewhat limited due to its limited range of dilution as well as the variable affinities for different volatiles found in its stationary phases. Thermal desorption (TD) is a more robust technique since it is not as selective, has a much larger capacity and samples rapidly. Although it is mostly used for GC/MS and single sniff analysis it is adaptable to dilution analysis. The incorporation of selected ion chemical ionization during GC-MS will allow the flavor chemist to identify and quantify compounds by the same technique used for GC/O.

In the end the challenge will be to determine how individual components function in mixtures to produce the entire sensory "movie". As a first step the use of n^{-1} descriptive analysis will indicate which compounds contribute to what olfactory perception. N^{-1} experiments are conducted with artificial flavors in a matrix designed to mimic a product. The descriptive analysis of the reference sample is compared to the results from samples that have one of the n components missing. From the results of each of the n elimination experiments the role of each component in the mixture can be estimated. The analysis of Linden honey (21) is an example of this approach (22).

Difference testing of natural and simulated samples, as well as in-depth descriptive analysis of n-dimensional component mixtures will be valuable in the correlation of chemical data (stimuli) and sensory data (response). These studies might also help in understanding the suppression and enhancement effects of certain combinations of compounds.

One of the newest applications of GC/O is its use to assess differences in individual olfactory acuity. Fig. 4 shows the odor spectra for two individuals using a set of standard compounds. The spectrum on the top shows the response of the reference individual to whom all the component concentrations have been adjusted to yield the same OSV:100. The spectrum below shows another individual with a distinct specific anosmia for one compound and a hyposmia for the compound 381. Using OSV's collected from individuals and compared to the genetic pattern produced from code for their olfactory receptor proteins could lead to an identification of the sensory function of different genes. In addition, the standard set

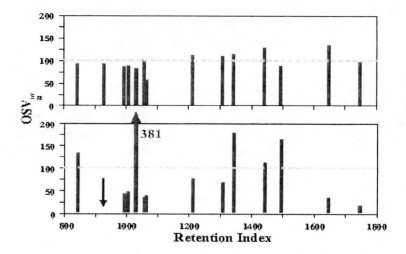

Figure 4. The odor spectrum of a set of standard odorants showing the specific anosmia of the subject shown in the lower spectrum compared to the reference individual shown above (23).

could be used to screen the subjects in sensory experiments for their sensory acuity and specific anosmias.

Sampling techniques that account for flavor release coupled with new selective and sensitive GC/MS equipment can now be used to determine most of the odor-active components in food needed to predict flavor. In addition, the development of on-line databases with a wealth of GC/O- derived organoleptic and chemical information puts a powerful pallet of tools at the disposal of the current-day flavor chemist. Despite these advances however, there continues to be a subset of odorants, though it is shrinking yearly, that is important to flavors and fragrances and that remain unknown. In the search for new and exciting flavorants, GC/O remains an important tool as more and more exotic plant materials are investigated as sources of flavorings for use in processed food systems. With the complexity of food matrixes a constant, GC/O methods provide invaluable information to chemists allowing them to focus their attention and resources on chromatographic areas of odor activity.

Literature Cited

1. Smith, R. M.S. Thesis, Cornell University, 1998
2. James, A.T. and Martin, A.J.P., *Analyst* 77, 915-932 1952
3. Fuller, G. H.; Steltenkamp, R.; Tisserand, G. A. The Gas Chromatography with Human Sensor: Perfumer Model. *Annals N. Y. Acad. Sci.*, 1964, *116*, 711-724.
4. Dravnieks and O'Donnell Principles and some techniques of high resolution headspace analysis. *J. Agric. Food. Chem.* 1971, *19*, 1049-1056.

5. Acree, T. E.; Butts, R. M.; Nelson, R., R.; Lee, C. Y.; Sniffer to determine the odor of gas chromatographic effluents. *Annal. Chem.* 1976, *48*, 1821-1822.
6. Acree, T. E. ; Barnard, J.; Cummingham, D. A procedure for the sensory analysis of Gas chromatography effluents. Food Chem. 1984, 14, 273-286.
7. Ullrich, F.;Grosch, W.Z. Lebensm. Unters. Forsch. 1987, 184, 277-282
8. Arn, H.; Acree, T.E. Flavornet: A Database of Aroma Compounds Based on the Odor Potency in Natural Products. *In Food Flavors: Formation, Analysis and Packaging Influences*, Elsevier: Lemnos, 1998.
9. Rothe, M.;Thomas B.Z Lebensm. Unters. Forsch. 1963, 119, 302-310.
10. Milo, Christian; Grosch, Werner. Detection of Odor Defects in Boiled Cod and Trout by Gas Chromatography-Olfactometry of Headspace Samples. J. Agric. Food Chem. (1995), 43(2), 459-62.
11. Roberts, D. D.; Acree, T. E. Simulation of Retronasal Aroma Using a Modified Headspace Technique: Investigating the Effects of Saliva, Temperature, Shearing, and Oil on Flavor Release. *J. Agric. Food Chem.* 1995, *43*, 2179–2186.
12. Schieberle, P. New developments in methods for analysis of volatile flavor compounds and their precursors. In *Characterization of Food: Emerging Methods*: Gaonkar, A.G., Ed., Elseiver: Amsterdam, 1995: pp 403.
13. Roberts, Deborah D.; Pollien, Philippe. Relationship between aroma compounds' partitioning constants and release during microwave heating. ACS Symp. Ser. (1998), 705(Flavor Analysis), 61-68.
14. Deibler, K.D., Acree, T.E., and Lavin, E.H., Solid Phase Microextraction application in Gas Chromatography-Olfactometry Dilution Analysis. *J. Agric. Food Chem.*, 2000, in Press.
15. Da Silva, Maria A.A.P.; Lundahl, David S.; McDaniels, Mina R. The capability and Psychophysics of Osme: a new GC-Olfactometry technique. Dev. Food Sci. (1994), 35 (Trends in Flavour Reseatch), 191-209. 1994
16. Rouseff, R. L.; Lee, W. E., III; Huefner, C. A. Time-intensity studies of citrus bitter compounds. Dev. Food Sci. (1990), 24(Flavors Off-Flavors '89), 213-23.
17. Stevens, S. S. The psychophysics of sensory functions. *Am. Scientist*, 1960, *48*, 226-253.
18. Acree, T.E. Bioassays in Flavor Research in *Flavor Science: Sensible Principles and Techniques*, Acree, T. E.; Teranishi, R., Eds.; American Chemical Society: Washington DC, 1993: pp 1-22.
19. Roberts, D.D.; Acree, T.E. Model Development for Flavour Release From Homogeneous Phases. In *Flavour Science Recent Developments*, Taylor,A. J.; Mottram, D. S., Eds.; The Royal Society of Chemistry: Cambridge, UK, 1996: pp 399-404.
20. ASTM (1979) Standard Practice for Determination of Odor and Taste Thresholds By a Forced-Choice Ascending Concentration Series Methods of Limits. *ASTM Spec. Tech. Pub.*, E 679, American Society for Testing Materials, Philadelphia, PA.
21. Blank, Imre; Fischer, Karl Heinz; Grosch, Werner. *Intensive neutral odorants of linden honey. Differences from honeys of other botanical origin.* Z. Lebensm.-Unters. Forsch. (1989), 189(5), 426-33.

22. Lawless, H. T.; Heyman, H. *Sensory Evaluation of Food: Principles and Practices*; International Thomson Publishing: New York, NY 1998.

23. Friedrich, J.E. and Acree, T.E., Issues in Gas Chromatography - Olfactometry (GC/O) Methodologies. In *Advances In Flavor Analysis*, Risch, S., Ed.; American Chemical Society: Washington DC, 2000.Abbott, N.; Etievant, P.; Issanchou, S.; Langlois, D.; Critical Evaluation of Two Commonly use techniques for the Treatment of Data from Extraction Dilution Sniffing Analysis. *J. Agric. Food Chem.* 1993, *41*, 1698-1703.

24. Abbott, N.; Etievant, P.; Issanchou, S.; Langlois, D.; Critical Evaluation of Two Commonly use techniques for the Treatment of Data from Extraction Dilution Sniffing Analysis. *J. Agric. Food Chem.* 1993, *41*, 1698-1703.

25. Acree, T.E.; Barnard, J., Gas Chromatography-Olfactometry using CharmAnalysis. In *Trends in flavour research*; Maarse, H., Ed.; Proceedings of the 7th Weurman Flavour Research Symposium; Elsevier: Amsterdam, 1994: Vol. 35, pp 211-220.

26. Acree, T.E. Flavor characterization. *Kirk Othmer Encycl. Chem. Technol.*, 1993, *4th Ed 11*, 1-16.

27. Acree, T.E. Gas Chromatography - Olfactometry. In *Flavor Measurement*. C.T. Ho, C. T.; Manley, C., Eds.; Marcel Dekker, Inc., New York, NY, 1993: pp 77-94.

28. Acree, T.E. GC/Olfactometry: GC with a sense of smell. *Anal. Chem.* 1997, *69*, 170 A-175 A.

29. Drawert, F.; Christoph, N. Significance of the sniffing-technique for the determination of odor thresholds of aroma impacts of trace volatiles. *In Analysis of Volatiles: Methods and Applications*, Schreier, P., Ed.; De Gruyter: Berlin, 1984: pp 269-291.

30. *Flavornet. Gas Chromatography-olfactometry (GC-O) of natural products*, URL http://www.nyseaes.cornell.edu/fst/faculty/acree/flavornet

31. Hinterholzzer, A.; Schiberle, P. Identification of the Most Odour-active Volatiles in Fresh, Hand-extracted Juiceof Valencia Late Oranges by Odour Dilution Techniques. *Flavour Fragr. J.* 1998, *13*, 49-55.

32. Marin, A. B.; Acree, T.E.; Barnard, J.; Variation in odor detection thresholds determined by charm analysis. *Chemical Senses*, 1988, *13*, 435-455.

33. Marin, A.B.; Barnard, J.; Darlington, R. B; Acree, T. E.; Sensory Thresholds: Estimation from Dose-Response Curves. *J. Sens. Stud.*, 1991, 205-225.

34. Nawar, W.W. Some Variables Affecting Composition of Headspace. *J. Agric. Food Chem.* 1971, *19*, 1057-1059.

35. Roberts, D.D.; Acree, T.E. Effects of Heating and Cream Addition on Fresh Raspberry Aroma Using a Retronasal Aroma Simulator and Gas Chromatography. *J Agric. Food Chem.*, 1996, *44*, 3919-3925.

36. Roberts, D.D.; Acree, T.E. Retronasal Flavor Release in Oil and Water Model Systems with an Evaluation of Volatility Predictors. In *Food Flavor Interactions*; McGorrin, R. and Leland, J., Ed., American Chemical Society: Washington DC, 1996: ACS Symposium Series, 633: pp 179-187.

Chapter 2

Application of Gas Chromatography–Olfactometry to Flavor Creation

Hisakatsu Iwabuchi, Yuriko Imayoshi, Yasue Yoshida, and Hiroshi Saeki

San-Ei Gen F.F.I., Inc., 1–11, Sanwa-cho 1 chome, Toyonaka, Osaka 561–8588, Japan

The combination of gas chromatography-olfactometry (GC-O) with sensitive instrumental methods, such as gas chromatography-mass spectrometry to analyze foods provides a powerful new approach to flavor development. Working with flavorist and customers to find aroma components that contribute to the concept of a new flavor, analytical chemists can assist flavorists in their development efforts. For example, using the GC-O method, CharmAnalysis, we have identified some new aliphatic esters, lactones and other compounds in the volatiles of white peach, grape (Kyohou), and pear (La France) that were used to build new flavor recipes. This paper will review the application of GC-O to flavor creation and report some of the structures of key aroma compounds found in foods during the development of new products.

The analysis methods of volatiles of natural products and foods has been advanced by many flavor scientists over the past 30 years, and their results have contributed to the development of flavor industries. In general, much evaluation experience and knowledge about many flavor materials along with a keen sense about flavor properties, blending flavor materials, application of flavors to foods and evaluation of their flavor properties are necessary for flavor creation. On the other hand, what's needed is chemical and biological knowledge for extraction of volatiles from materials and instrumental analysis to elucidate their constituents for flavor research. So far, a tremendous amount of chemical information, obtained mainly from Gas Chromatography-Mass spectrometry (GC-MS), has been used for flavor development. However, it has been pointed out that it is difficult not only to build flavors, but also to represent original aroma by using only such chemical information.

Recently, the usefulness of Gas Chromatography-Olfactometry (GC-O) has been revalued and it has evolved as an analytical method mainly for finding characterizing odorants in natural products and foods. However, it seems that the use of the GC-O is limited to flavor chemists and sensory scientists due to its complicated operation and relatively low reproducibility of evaluations.

On the other hand, flavorists tend to make subjective evaluations about various aromas. But, they can express complex flavor properties in plain words that people can understand easily. Also, they have been trained to combine various flavor chemicals and their flavor properties. Thus, it is suggested that the use of GC-O can be augmented by utilizing the abilities of flavorists. Furthermore, working with flavorists and customers to find aroma components by using GC-O as an interface method that contributes to the concept of a new flavor, flavor chemists can assist flavorists in their development efforts to save time.

Experimental Procedure

Investigated Fruits. The fruits analyzed in this study are as follows :
1. White peach (*Rosaceae, Prunus persica* `Hakuhou') harvested in 1998, Wakayama.
2. Grape (*Vitaceae, Vitis vinifera* `Kyohou') harvested in 1998, Yamanashi.
3. Pear (*Rosaceae, Pyrus communis* var. *sativa* `La France') harvested in 1998, Yamagata.

Sample Preparation. For the isolation of volatiles of the fruits three different procedures were employed. These were the Dynamic Headspace Method (DHS), the Reduced Pressure Steam Distillation Method (RSD), and the Liquid-Liquid Extraction Method.

Dynamic Headspace Method (DHS). Intact fruits (total weight 12.5kg) were placed in a 30-L stainless container. A Pyrex glass head attached to the top of the container which allowed purified air to enter the bottom of the container and exit through a Tenax trap, consisting of a glass tube 20 cm x 1.0 cm (i.d.) filled with 1.0 g of Tenax (35/60 mesh, GL Science, Ltd.; Tokyo). Sampling was continued at room temperature for 12 h at 300 ml/min. The collected volatiles were eluted from the trap with freshly distilled diethyl ether (10 ml) and carefully concentrated with a Vigreux column to a final volume of ca. 0.5 ml.

Reduced Pressure Steam Distillation Method (RSD). The fleshy part of the fruits (400 g, core and seeds were excluded) was blended with 200 ml of 30% calcium chloride solution in a Waring blender for 10 s. Four batches were prepared using a total of 2.0 kg of flesh. After removing the residue by centrifugation (6000 rpm, 5 °C, 30 min), the resulting solution was added to a 10 L round-bottomed flask. An additional 1.5 L of distilled water was added to the flask. RSD was performed (10 torr, 40 °C) for 4 hr and 1.0 L of the distillate was collected. The low boiling volatiles were collected in the cooling traps with dry ice-acetone and liquid nitrogen). The distillate and the contents of the traps were then combined and were extracted with an organic solvent (freshly distilled diethyl ether or dichloromethane). The extract was carefully concentrated with a Vigreux column to a final volume of ca.1 ml.

Liquid-Liquid Extraction Method. The fleshy part of the fruits (2 kg) was blended with 500 ml of 30% calcium chloride solution in a Waring blender for 10 s. After removing the residue by centrifugation (6000 rpm, 5 °C, 30 min), the resulting solution was added to a 5 L round-bottomed flask, and the solvent (dichloromethane, 700 ml) was added slowly. The mixture was stirred gently with a magnetic stirrer for 1 hr at room temperature, then separated in separatory funnel, and dried over anhydrous sodium sulfate. The extract was carefully concentrated in a rotary evaporator to a final volume of ca.3 ml.

Instrumental Analysis. General: Nuclear Magnetic Resonance (NMR) spectra were taken on JEOL GSX500 and Lambda 400 instruments (Akishima, Tokyo) in $CDCl_3$ with tetramethyl silane as internal standard. Infrared (IR) spectra were recorded on a JASCO FTIR-350 (Hachioji, Tokyo). Optical rotation was recorded on a JASCO DIP-1000 and Circular Dichroism (CD) spectrum was recorded on a JASCO J-720W in methanol solution. For preparative high performance liquid chromatography (HPLC) a Chemcosorb Si-10 (30 cm x 10 mm i.d., Chemco, Inc., Osaka) was used.

GC-MS: A Hewlett-Packard (Avondale, PA) 5973 mass selective detector equipped with a 0.25 mm x 60 m fused silica capillary column coated with DB-WAX (J&W Scientific, Folsom, CA) was utilized for all GC-MS analysis. The column was held at 50 °C for 2 min and then programmed at 3 °C/min to 220 °C which was held for 60 min. The injector temperature was 250 °C, and the ion source was held at 230 °C. Helium carrier gas was used at a column head pressure of 25 psi. The instrument was operated in the electron impact mode at 70 eV, taking scans from 30 to 300 m/z. The separated compounds were identified by matching their mass spectra and Kovats indices with those of reference standards. In some cases reference compounds were synthesized in our laboratory.

Gas Chromatography-Olfactometry: Extracted samples were analyzed using a GC-O system (CharmAnalysis(*1*)) on a modified HP 6890 GC (Datu, Inc., Geneva, NY) and sniffed using a 0.32 mm x 15 m fused silica capillary column coated with DB-WAX. The oven temperature was programmed from 40 to 220 °C at 5 °C/min. The retention time of each odorant was converted to Kovats indices using 10-32 - carbon normal paraffins. All sniffing samples were serially diluted by 3-fold. The odor was sniffed by two or three experienced members.

Chiral GC Analysis: A HP 5890 Series II GC and GC-O system equipped with a 0.25 mm x 20 m Chiraldex G-TA (Advanced Separation Technologies Inc., NJ) was used for chiral GC analysis. The column was held at 110 °C and then programmed at 2 °C/min to 135 °C and held there for 30 min.

Isolation of 3-methyl-(*E*)-5-decen-4-olide: 3-methyl-(*E*)-5-decen-4-olide was isolated from another extract of white peach. The slurry prepared from the fleshy part of the white peach (10 kg, core and seeds were excluded) and 10 L of water was added to a 30 L stainless container and steam distillation was performed. The resulting distillate (5 L) was saturated with sodium chloride and extracted with dichloromethane (1 L). Five batches were carried out and finally 43 g of the extract was obtained. The pure 3-methyl-(*E*)-5-decen-4-olide was isolated by chromatography on a silica gel column eluted with n-hexane-ether (4:1,v/v) and subsequent preparative HPLC (n-hexane:isopropanol=95-97:5-3,v/v). Spectral data of this compound were as follows: $[\alpha]_D$ -107.5 (*c* 0.07, MeOH), UV: λ_{max} 210 nm (MeOH), MS: *m/z* 182 (M$^+$), 125, 113, 95, 69, 57, 41. IR : 1760 cm^{-1} (CHCl$_3$). ^{1}H-NMR (500 MHz, CDCl$_3$, ppm, *J* in Hz): 0.90(3H, t, *J*=7.20), 1.12 (3H, d, *J*=6.35), 1.35 (4H, m), 2.09 (2H, m), 2.20 (1H, dd, *J*=10.50, 16.36), 2.25 (1H, m), 2.67 (1H, dd, *J*=7.08, 16.35), 4.35 (1H, t, *J*=7.81), 5.44 (1H, ddt, *J*=15.38, 7.81, 1.46), 5.82 (1H, dt, *J*=15.38, 6.47). ^{13}C-NMR (125 MHz, CDCl$_3$, ppm): 13.72, 16.15, 21.98, 30.83, 31.70, 36.76, 37.03, 87.74, 126.16, 136.70, 176.30.

Results and Discussion

White peach. Peach has a very characteristic flavor that is mainly composed of fruity, green and milky notes. According to the classical literature, peach seems to have already been a popular fruit about 1250 years ago in Japan. Today, many varieties of peach are cultivated in Japan and its fresh fruit is a popular summer fruit. Beside fresh eating, various kinds of foods and beverages with peach flavors are sold in great volume in the Japanese market. The main purpose of this study is to find useful flavor components for creating flavors matched with these processed foods. In this study, we used Hakuhou (*Rosaceae, Prunus persica* 'Hakuhou'), which was the most popular white peach cultivated in Wakayama Prefecture, located in central Japan.

The volatiles of white peach were isolated by using the dynamic headspace method (DHS) of intact fruit and the reduced pressure steam distillation method. The extracts were analyzed by GC-MS and many volatile components were identified which were useful information for building up the framework of peach flavors. Figure 1 shows typical total ion chromatograms (TIC) of these volatiles. It is well known that the components identified by conventional GC-MS alone can' t represent the original aroma of targets in many cases. Therefore, the GC-O method was employed to find hidden odor active components to prepare new peach flavor formulae.

Figure 2 shows the gas and aroma chromatograms of the volatiles obtained by the RSD method. This GC-O result showed that the main components such as aliphatic alcohols and esters did not contribute to peach aroma as key components. The peak 1 in Figure 2, one of the characterizing odorants, possessing green and hay notes, was identified to be (Z)-2-hexenol and evaluated as an undesirable note for peach flavor. On the other hand, the peaks 4 and 7 in Figure 2 were evaluated as useful components for peach flavors. One of these potential components (peak 7 in Figure 2) was easily identified to be (Z)-6-dodecen-4-olide.

Another characterizing sweet and green odorant (peak 4) in Figure 2 was hidden under other GC peaks, and had to be isolated from other volatiles obtained from white peach by using silica gel column chromatography and preparative HPLC. The molecular ion peak of peak 4 was observed at m/z 182 (Figure 3) and the NMR spectrum showed the presence of characteristic protons as follows: a primary methyl proton at δ 0.90, a secondary methyl proton at δ 1.12, an allylic methylene proton at δ 2.09, a methylene proton adjacent to carbonyl group at δ 2.20 and 2.67, a methine proton adjacent to oxygen at δ 4.35, two olefinic protons at δ 5.44 and 5.82. IR spectrum (1760 cm^{-1}) and ^{13}C-NMR spectrum (δ 176.30) indicated the presence of γ-lactone moiety. The coupling constant (15.38 Hz) of two olefinic protons indicated the *trans* geometry of the double bond. To make the relative structure clear, decoupling and nuclear Overhauser effect (nOe) experiments were employed. Irradiation at δ 4.35 (a methine proton adjacent to oxygen) affected an olefinic proton signal (δ 5.44) and a methine proton signal (δ 2.25, m). This fact showed the double bond was attached to a methine carbon adjacent to oxygen. The nOe effects were observed at a methine proton signal adjacent to oxygen (δ 4.35) and a methylene proton adjacent to carbonyl group (δ 2.26) by irradiation at δ 1.12 (a secondary methyl proton). This fact indicated that the secondary methyl group attached at C3 position of the γ-lactone ring and the secondary methyl group and unsaturated side chain were expressed by an *anti* configuration. Thus, the relative structure of peak 4 in Figure 2 has been identified as 3-methyl-(E)-5-decen-4-olide, whose structure was finally established by synthesis in our laboratory (3). To elucidate the absolute structure of peak 4, the CD spectrum was measured and it showed a (+)-Cotton effect. On the basis of the perspective view, peak 4 was identified as (3S, 4R)-(-)-3-methyl-(E)-5-decen-4-olide.

In addition, the chirospecific GC-O performed using synthetic 3-methyl-(E)-5-decen-4-olide revealed that the (-)-enantiomer, the natural type, had more characteristic sweet, fresh and juicy aroma and had a lower odor threshold than the (+)-enantiomer. This suggests that enantiomeric lactone could be used to make characteristic new peach flavors.

(Z)-6-dodecen-4-olide   (3S, 4R)-(-)-3-methyl-(E)-5-decen-4-olide

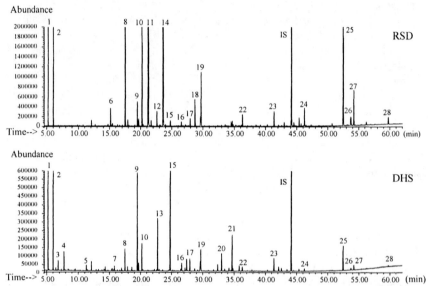

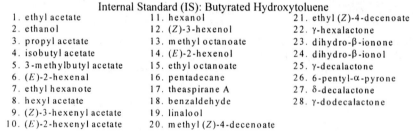

Figure 1. Typical Total Ion Chromatograms of the Volatiles of White Peach
Internal Standard (IS): Butyrated Hydroxytoluene

1. ethyl acetate
2. ethanol
3. propyl acetate
4. isobutyl acetate
5. 3-methylbutyl acetate
6. (E)-2-hexenal
7. ethyl hexanote
8. hexyl acetate
9. (Z)-3-hexenyl acetate
10. (E)-2-hexenyl acetate

11. hexanol
12. (Z)-3-hexenol
13. methyl octanoate
14. (E)-2-hexenol
15. ethyl octanoate
16. pentadecane
17. theaspirane A
18. benzaldehyde
19. linalool
20. methyl (Z)-4-decenoate

21. ethyl (Z)-4-decenoate
22. γ-hexalactone
23. dihydro-β-ionone
24. dihydro-β-ionol
25. γ-decalactone
26. 6-pentyl-α-pyrone
27. δ-decalactone
28. γ-dodecalactone

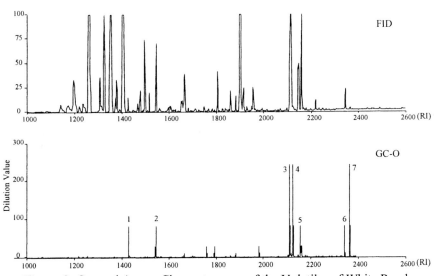

Figure 2. Gas and Aroma Chromatograms of the Volatiles of White Peach

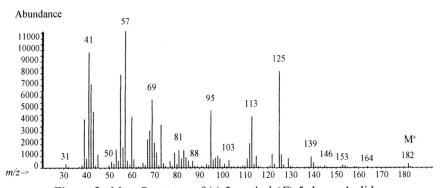

Figure 3. Mass Spectrum of (-)-3-methyl-(*E*)-5-decen-4-olide

Grape. Grape is a popular flavor in the Japanese food market, especially in carbonated beverages and desserts. Today, many varieties of grape are cultivated in Japan, of which Kyohou (*Vitaceae, Vitis vinifera* 'Kyohou') has been called King of the grapes. This variety, created as a hybrid of Campbell early and Centennial, was named Kyohou in 1945. Its flavor properties are characteristic sweet, fruity and foxy notes. The volatiles were isolated from the fleshy part of Kyohou by using the liquid-liquid extraction method with dichloromethane in this study. Figure 4 shows the gas and aroma chromatograms of the volatiles of the flesh part of Kyohou. Among 8 characterizing odorant peaks, the peak No.7, possessing a powerful grape, foxy and sweet note, was evaluated as the most interesting and important flavor component. This peak No.7 on the aroma chromatogram corresponded to a small peak on the total ion chromatogram and was identified as *o*-aminoacetophenone by comparing it with an authentic sample. The region of interest in the total ion chromatogram of the volatiles of Kyohou and the mass spectrum corresponding to odor active peak No.7 in Figure 4 is shown in Figure 5. This compound has been reported as a foxy smelling component of the *Vitis labruscana* species (*4*).

Table I summarizes the results of GC-O on Kyohou, in which peaks 2 and 3 were unknown and peak 6 was identified as 2,5-dimethyl-4-hydroxy-2,3-dihydro-3(2*H*)-furanone. These compounds were found to be undesirable notes for flavor creation by our flavorist and the customer in this study. On the other hand, four categories of flavor properties in Table I, that is, green and fresh, honey and sweet, and fruity and foxy notes are essential for flavor creation. Methyl anthranilate and its derivatives such as ethyl anthranilate and dimethyl anthranilate have been used for grape flavors as key aroma chemicals. Among them, methyl anthranilate is preferred for general use. Indeed, methyl anthranilate is a nature identical component found in the *Vitis labruscana* species (*5*), and this compound has a foxy note, but also a slightly burnt note. Methyl anthranilate was not found in the volatiles of Kyohou, but *o*-aminoacetophenone was. Our finding of this compound is very useful chemical information for building up a new type of grape flavor possessing sweet and foxy notes.

Table I. Summarized Aroma Impression of the Volatiles of Kyohou

Desirable Notes			Undesirable Notes		
Descriptors	Components	peaks in Figure 4	Descriptors	Components	peaks in Figure 4
Green, Fresh	(*E*)-2-heptenal	1	Aldehydic, Fishy	unknown	2
	(*E*)-2-hexenoic acid	5	Hay	unknown	3
Honey	phenethyl alcohol	4	Caramel, Burnt	2,5-dimethyl-4-hydroxy-3(2H)-furanone	6
Sweet	phenylacetic acid	8			
Fruity, Foxy	*o*-aminoacetophenone	7			

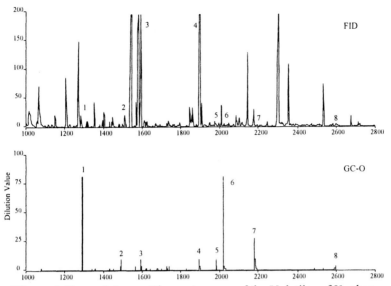

Figure 4. Gas and Aroma Chromatograms of the Volatiles of Kyohou

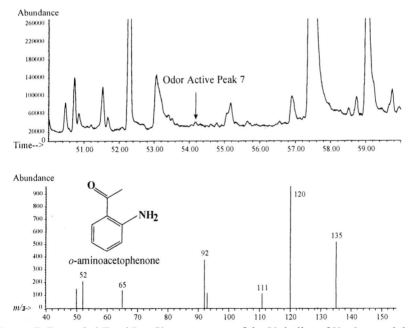

Figure 5. Expanded Total Ion Chromatogram of the Volatiles of Kyohou and the Mass Spectrum of Peak 7 in Figure 4

Pear. The main purpose of this part of the study is to find odor active components of La France pear (*Rosaceae, Pyrus communis* var. *sativa* 'La France'). Pear varieties are classified into three groups: European-type pears (*Pyrus communis* L.) such as Bartlett, La France and Passe Crassane, Asian pears (*Pyrus pyrifolia* (Burm.f.) 'Nakai) such as Nijyusseiki (20th century), Kousui and Housui, and Chinese pears (*Pyrus ussuriensis* Maxim.) such as Yali and Tuli.

La France pear is a typical popular autumn fruit in Japan beside Asian pears, and the producing district of this pear is Yamagata prefecture located in north Japan. Shiota reported the volatiles, isolated by the simultaneous distillation and extraction method, and changes in the composition of the volatiles during maturation of La France pear (*6*). In this study, we carefully isolated the volatiles from mature fruits using the RSD method to investigate further odor active components of this fruit.

Figure 6 shows a typical total ion chromatogram of the volatiles of La France pear. The major components of the volatiles were aliphatic alcohols, such as propanol, isobutanol, butanol and hexanol, and aliphatic esters, mainly acetates of aliphatic alcohols, such as butyl acetate, pentyl acetate and hexyl acetate. Our own GC-O result (Figure 7) showed that trace volatile components were the most significant contributors to characteristic pear notes, rather than major components such as aliphatic alcohols and esters. Especially, the two components corresponding to peaks 2 (Charm value (*7*): 1641, fruity-green and pear-like odor) and 3 (Charm value: 688, pear-like odor) in Figure 7 were essential to developing pear flavors. Other characterizing odorants were identified as 2,4-hexadienyl acetate (peak 1, Charm value: 1485, green, waxy, and hay odor), beta-phenethyl acetate (peak 4, Charm value: 847, honey and pleasant odor) and ethyl cinnamate (peak 6, Charm value: 809, honey and pleasant odor) by comparing them with authentic samples, respectively. The most characterizing odorants, peaks 2 and 3, were identified as (*Z*)-5- and (*E*)-2-octenyl acetates, respectively, by comparing them with authentic samples. These esters have also been found in banana (*8,9*), but are reported here the first time in La France pear. (*E*)-2-octenyl acetate was previously identified in the volatiles of Chinese pear, Yali (*10*). Interestingly, this ester has also been recently reported as a sex pheromone of the plant bug (*11*). Also, (*Z*)-5-octenyl acetate has been recently reported in yellow passion fruit (*12*). Furthermore, ethyl (*E,Z*)-2,4-decadienoate (*13*), which is called pear ester and has been reported as the character impact compound of Bartlett pear (*14*), was evaluated as having weak green and fruity notes (peak 5, Charm value: 79) as shown in Figure 7. This result shows that pear ester, which has been considered as an important flavor material for pear flavors, is not necessarily important for the flavor creation of La France-type pear flavors. And, our findings of (*E*)-2 and (*Z*)-5-octenyl acetates contributed to building new flavor recipes.

Conclusions

The possibility of using GC-O as a method for flavor creation was studied through our actual fruit flavor developments and its usefulness was recognized by us. Many flavor researchers have pointed out the many aspects of GC-O and many

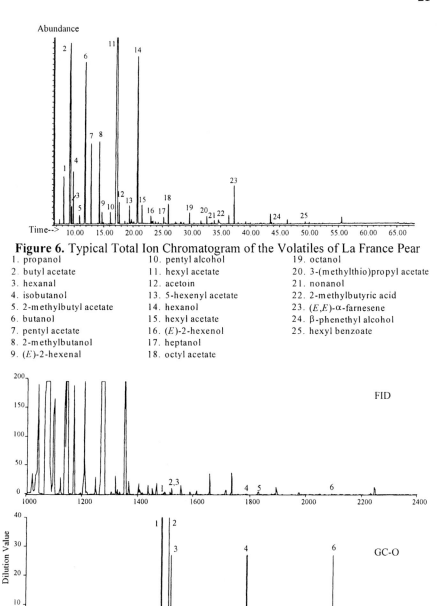

Figure 6. Typical Total Ion Chromatogram of the Volatiles of La France Pear

1. propanol
2. butyl acetate
3. hexanal
4. isobutanol
5. 2-methylbutyl acetate
6. butanol
7. pentyl acetate
8. 2-methylbutanol
9. (*E*)-2-hexenal

10. pentyl alcohol
11. hexyl acetate
12. acetoin
13. 5-hexenyl acetate
14. hexanol
15. hexyl acetate
16. (*E*)-2-hexenol
17. heptanol
18. octyl acetate

19. octanol
20. 3-(methylthio)propyl acetate
21. nonanol
22. 2-methylbutyric acid
23. (*E,E*)-α-farnesene
24. β-phenethyl alcohol
25. hexyl benzoate

Figure 7. Gas and Aroma Chromatograms of the Volatiles of La France Pear

excellent solutions have been proposed, for example Stable Isotope Dilution Assay and the concept of Odor Activity Value. However, it is no exaggeration to say that there is no single analytical and evaluation technique for flavor research. Rather, flavor research and development must be carried out by making use of the advantages of each of the analytical methods in combination. When applying GC-O to flavor creation, the most important point we consider is that GC-O is not only an analytical method for finding odor active components and evaluating flavor properties, but is also a technology which allows flavor chemists, flavorists and even customers to share flavor impressions on a time-axis basis, and to develop mutual understanding of flavor properties and demands. This enables the successful development of the flavors which meet the customer needs, while at the same time producing data to advance academic studies supporting flavor development.

Acknowledgments

We are grateful to Dr. Terry E. Acree of Cornell University for his helpful discussion and Dr. Nobuo Kato of Kyushu University for performing the NMR measurements of 3-methyl-(*E*)-5-decen-4-olide.

References and Note

1. Acree, T. E.; Barnard, J.; Cunningham, D. *Food Chem.* **1984**, *14*, 273-286.
2. Boldingh, J.; Taylor, R. J. *Nature* **1962**, *194*, 909-913.
3. Iwabuchi, H.; Yukawa, C. *Jpn. Kokai Tokkyo Koho* JP09087267;
   CA 127(2)17581d.
4. Acree, T. E.; Lavin, E. H.; Nishida, R.; Watanabe, S. In *Flavour Science And Technology, 6th Weurman Symposium, Geneva, Switzerland*; Bessiere, Y Thomas, A. F., Eds.; Wiley: New York, **1990**; pp 49-52.
5. Robinson, W. B.; Shaulis, N. J.; Pederson, C. S. *Fruit Prod. J. Am. Food Manuf.* **1949**, *29*, 36-37, 54, 62: Fuleki, T. *Can. J. Plant Sci.* **1972**, *52*, 863-868.
6. Shiota, H. *J. Sci. Food Agric.* **1990**, *52*, 421-429.
7. Acree, T. A. *Anal. Chem. News Features* **1997**, *69*, 170A-175A.
8. Quast, S. V. *Cienc. Cult. (Sao Paulo)* **1976**, *28*, 348-352.
9. Berger, R. G.; Drawert, F.; Kollmannsberger, H. *Chem., Mikrobiol., Technol. Lebensm.* **1986**, *10*, 120-124.
10. Kawabata, N.; Imayoshi, Y.; Iwabuchi, H. *Foods Food Ingredients J. Jpn.* **1994**, *160*, 93-98.
11. Millar, J. G.; Rice, R. E. *J. Econ. Entomol.* **1998**, *91*, 132-137: Millar, J. G.; Rice, R. E.; Wang, Q. *J. Chem. Ecol.* **1997**, *23*, 1743-1754.
12. Werkhoff, P.; Guentert, M.; Krammer, G.; Sommer, H.; Kaulen, J. *J. Agric. Food Chem.* **1998**, *46*, 1076-1093.
13. Shiota (*6*) reported that ethyl (*E,Z*)-2,4-decadienoate was not found in the volatiles of La France pear but the corresponding *E,E*-isomer was found.
    In our study, we couldn't confirm the presence of the *E,E*-isomer.
14. Jennings, W. G.: Creveling, R. K.; Heinz, D. E. *J. Food Sci.* **1964**, *29*, 730-734.

Chapter 3

Characterization of the Odor-Active Constituents in Fresh and Processed Hops (variety *Spalter Select*)

Peter Schieberle and Martin Steinhaus

Institute of Food Chemistry, Technical University of Munich and German Research Center for Food Chemistry, Lichtenbergstrasse 4, D–85748 Garching, Germany

By application of dilution to odor threshold techniques, trans 4,5-epoxy-(E)-2-decenal (metallic), linalool (flowery) and myrcene (geranium-like) were characterized as important odorants in hops. Among the 18 aroma-active compounds showing Flavor Dilution factors higher than 16, in particular, 4,5-epoxy-(E)-2-decenal, 1,(Z)5-octadien-3-one, 1,(E)3,(Z)5-undecatriene and 1,(E)3, (Z)5,9-undecatetraene were newly identified in the hop essential oil. Quantitative measurements performed by stable isotope dilution assays on nine odorants having FD-factors higher than 128 confirmed their flavor contribution, because they much exceeded their odor threshold. Hop drying led, in particular, to a degradation of (Z)-3-hexenal, identified as the most important odorant in fresh hop cones, and also the undecaenes.

Hops are used by the brewing industry, in particular, to render a bitter taste and a more or less pronounced "hoppy" aroma to beer. Another important factor is the content of phenolic compounds, such as tannins, which significantly influence the beer stability.

The hop plant *Humulus lupulus* is a climber reaching heights of more than 7 meters. The flowers or blossoms of the female plant are quite small and greenish-yellow in color. After ripening, these so-called cones are harvested and dried from an initial water content of about 80 % to about 12 %. All of the important chemicals, such as the bitter-tasting acids, the tannins and the aroma compounds are present in the so-called lupulin glands, a yellow powder which is mechanically isolated from the cones. The world crop of hops in 1998 was about 85000 tons and nearly 80 % of the total crop was grown in the U.S. and Germany (German Hop Growers Association, 1998). In Germany, the biggest hop growing area is the Hallertau where the predominant varieties *Perle*, *Magnum* or *Spalter Select* are grown.

The compounds responsible for the bitter taste, the so-called humulones and lupulones, are well characterized and amount to about 5 to 8 percent of the dry weight. The hop aroma is caused by components present in the essential oil which amounts to 0.5 to 3 % depending on the variety. However, compared to the bitter tasting compounds, the components responsible for the characteristic hop aroma are not yet fully understood.

Chapman (1, 2) and Semmler (3) have identified myrcene, linalool, humulene and 3-methylbutanoic acid as predominant constituents in the hop oil and have suggested that these are important odor contributors. Today, more than 400 volatiles are known to be present in different hop varieties (4).

The odor unit concept (OU; ratio of odorant concentration to odor threshold), also assigned as odor activity value (OAV), allows one to rank single odorants according to their odor contribution based on odor thresholds in the food matrix. Guadagni et al. (5) were the first to apply this concept on the aroma fraction of *Brewers Gold* hops. Their results, which are, in part, summarized in Table I, revealed 69 % of the overall odor intensity to be represented by the hydrocarbon fraction and 31 % by the oxygenated fraction. Myrcene was characterized to be by far the most potent individual contributor to the overall hop aroma. Methyl thiohexanoate and methyl decenoate were identified as further important odorants in the oxygenated fraction. However, a major part of the odor could not be assigned to single aroma compounds.

Table I. Results of the application of the odor unit (OU) concept to "*Brewers Gold*" hops (modified from (5))

Fraction	%	Odor threshold (µg/L water)	OU	% of total OU
Whole oil	100	12	83	100
Hydrocarbons	86	15	57	69
Myrcene	63	13	49	58
Humulene	15	120	1.3	1.5
Oxygenated fraction	14	5	28	31
Methyl thiohexanoate	0.12	0.3	4	4.8
Methyl decenoate	0.75	3	2.5	3.0

[a] Odor Units (OU) were calculated by dividing the concentrations in the hop oil (based on 1 mg of oil) by the respective odor thresholds in water.

The purpose of the following investigation was, therefore, to characterize the odor-active compounds in dried hop cones (variety: *Spalter Select*) by means of dilution to odor threshold techniques (cf. review by Schieberle, in (6)), to develop methods for their exact quantification and to gain first insights into the effect of drying on the overall hop aroma.

Experimental Part

Dried and fresh hop cones (crop 1996) were supplied by a local hop farmer. For GC/O-Headspace experiments, the dried hop cones (0.1 g) were placed in a septum-sealed, thermostated vessel (total volume: 50 mL) and equilibrated for 20 min. An aliquot (10 mL) of the headspace gas was withdrawn using a gas-tight syringe and analyzed by HRGC/Olfactometry as shown in Figure 1. For the dilution experiments decreasing volumes (10 mL, 5 mL, 2.5 mL .... 0.8 mL) were used which were taken separately from a freshly prepared sample of hops.

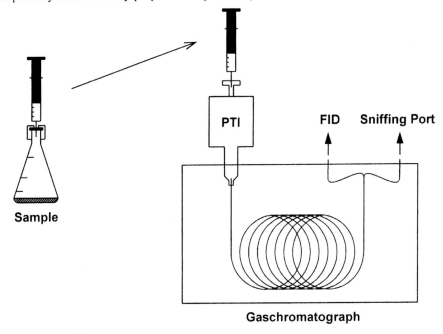

Figure 1. Scheme of the Static Headspace/Olfactometry

For the Aroma Extract Dilution Analysis, hops (5 g) were extracted with diethyl ether (total volume: 100 mL) and the volatile material was isolated by means of the SAFE-distillation head (8). The identification experiments were performed as described recently (7). Details on the quantification methods will be published elsewhere.

Results

Identification of key odorants in dried hops

Application of GC/O-Headspace on a sample of dried hop cones revealed 24 odor-active compounds (7). In Table II, the eight odorants appearing in the lowest headspace volumes used are displayed (1.25 mL down to 0.08 mL). In a headspace sample of 80 µL, only two odorants were detected and identified by HRGC/MS as myrcene with a geranium-like odor, and linalool exhibiting a flowery aroma. Increasing the volume to 630 µL allowed the detection of two further aroma compounds, besides myrcene and linalool: the fruity smelling ethyl 2-methylpropanoate and an unknown compound (no. 23) with a fresh, balsamic aroma.

Table II. Key odorants detected by GC/Olfactometry in dry hop cones - A total of 24 odorants was identified (7)

Volume (mL)[a]	Retention index[b]	Odor quality	Odorant[c]
0.08	988	geranium-like	Myrcene
0.08	1100	sweet, flowery	Linalool
0.63	754	fruity	Ethyl 2-methylpropanoate
0.63	**1174**	**balsamic, fresh**	1,(E)3,(Z)5-undecatriene and 1(E),3(Z)5,9-undecatetraene
1.25	772	fruity	Methyl 2-methylbutanoate
1.25	848	fruity	Ethyl 2-methylbutanoate
1.25	936	fruity	Unknown
1.25	969	rotten, sulfury	Dimethyltrisulfide

[a] Lowest headspace volume in which the odorant could be detected by HRGC/Olfactometry (0.1 g hops in a 50 mL vessel).
[b] Retention index on a Silicone SE-54 stationary phase.
[c] Identification is based on the use of reference odorants and a comparison of mass spectra, retention indices, odor quality and odor threshold.

For the identification of compound 23, 500 g of hops were extracted. The extract was purified by SAFE-distillation, followed by separation of the volatiles into a hydrocarbon and an oxygenated fraction by chromatography on silica gel (7). Separation of the hydrocarbon fraction by HRGC on a Silicone-54-fused silica column and sniffing of the effluent revealed a very small peak for compound 23 (Figure 2) which was further enriched by preparative GC and then separated on a Free Fatty Acid-Phase (FFAP) fused silica column. By sniffing of the effluent, two odor-active areas were detectable (Figure 3) with retention indices of 1387 (compound 23a) and 1448 (compound 23b). Based on syntheses of the reference compounds (7), 23a was identified as 1,3(E),5(Z)-undecatriene and compound 23b as 1,(E)3,(Z)5,9-undecatetraene (7). The structural characterization of this previously unknown undecatetraene will soon be reported in more detail (Steinhaus and Schieberle, J. Agric. Food Chem., submitted).

To gain a deeper insight into the composition of the key hop aroma compounds, a solvent extract was prepared of the hops (5 g) and purified by high vacuum distillation. Application of the Aroma Extract Dilution Analysis on the fraction of the neutral/basic volatiles revealed 34 odor-active regions in the Flavor Dilution (FD)-chromatogram (7) among which myrcene, linalool and 4,5-epoxy-(E)-

2-decenal, were the most odor-active. The results of the identification experiments in combination with the FD-factors are summarized in Table III. Besides linalool, myrcene, trans-4,5-epoxy-(E)-2-decenal and the two undecaenes, nonanal and the two fruity smelling esters ethyl 2-methylpropanoate and methyl 2-methylbutanoate were identified as further important hop aroma constituents. Eleven of the 18 odorants identified in the FD-factor region of 16 to 4096 (Table III) were previously unknown as hop constituents.

Quantitative measurements

In order to link data obtained by dilution to odor threshold techniques with the situation in the food itself, or to follow changes induced by manufacturing, exact quantitative data are a prerequisite. In particular, if trace aroma compounds have to be analyzed, a stable isotope dilution assay using 100 % isotopically labeled internal standards is the method of choice (6). Since most of these standards are commercially not available, synthetic routes for their synthesis have to be developed. In Figure 4 the synthetic approach used for $[^2H]_4$-1,3,5-undecatriene preparation is shown. The synthesis starts from 1-hexinol which is deuterated yielding $[^2H]_4$-hexanol. Oxidation of the alcohol with Dess-Martin-periodinane (9) gives the corresponding aldehyde which is finally condensed with 1,3-pentadienyltriphenyl phosphonium in a Wittig reaction to yield the (Z)-5- and (E)-5-isomer of 1,3(E)5-undecatriene. The mixture was finally separated into the isomers by Argentation Chromatography.

A total of 20 hop odorants was quantified by mass chromatography in combination with a stable isotope dilution assay (Steinhaus and Schieberle, J. Agric. Food. Chem., 2000, in preparation). The results obtained for nine odorants which have been identified with high FD-factors (cf. Table III) are displayed in Table IV. Myrcene and linalool showed by far the highest concentrations, whereas the newly identified hop odorants 4,5-epoxy-(E)-2-decenal, 1,(Z)5-octadien-3-one, 1,(E)3,(Z)5-undecatriene and 1,(E)3,(Z)5,9-undecatetraene occurred in ppb concentrations. However, a comparison with the odor thresholds in air confirmed the importance of the odorants under investigation for the overall hop aroma. Studies to mimick the hop aroma by recombination experiments are underway.

Differences between fresh and dried hop cones

Application of the Aroma Extract Dilution Analysis on a sample of fresh, undried hops of the same batch of *Spalter Select* hops revealed the same important odorants as identified in the dried sample. However, some odorants showed significant differences in their Flavour Dilution (FD)-factors which, with some limitations, are proportional to the respective concentrations. A comparison of the key odorants identified in the fresh sample (Figure 5) to those identified in the dried cones (Table IV) revealed that, in particular, (Z)-3-hexenal was much higher in the fresh hops, whereas the undecatriene and undecatetraene as well as 4,5-epoxy-(E)-2-decenal were somewhat lower.

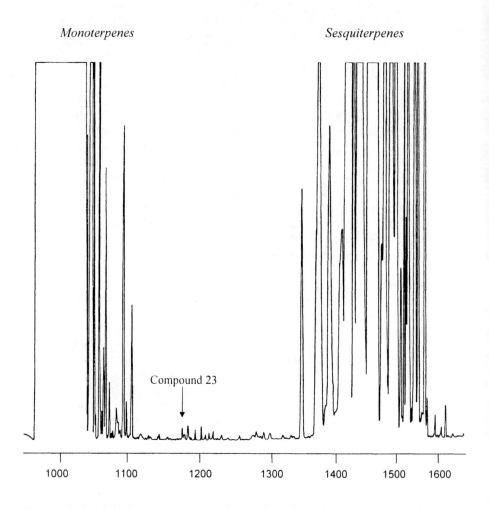

Figure 2. HRGC chromatogram of the hydrocarbon fraction isolated from hop cones

Table III. Key odorants (FD ≥16) in dried hop cones[a]

Odorant	FD-factor	Odorant	FD-factor
trans-4,5-Epoxy-(E)-2-decenal*	4096	2- and 3-Methylbutanoic acid	64
Linalool	2048	Propyl 2-methylbutanoate*	64
Myrcene	1024	Butanoic acid	64
Ethyl 2-methylpropanoate*	128	1-Octen-3-one*	32
Methyl 2-methylbutanoate*	128	4-Ethenyl-2-methoxyphenol*	32
Nonanal	128	Unknown (citrus-like)	32
1,(Z)5-Octadien-3-one*	128	Ethyl 2-methylbutanoate*	16
1,(E)3,(Z)5-Undecatriene*	128	Hexanal and (Z)-3-Hexenal*	16
1,(E)3,(Z)5,9-Undecatetraene*	128	Methional	16

[a] A total of 34 odor-active regions was detected.
* Previously unknown as volatile hop constituent.

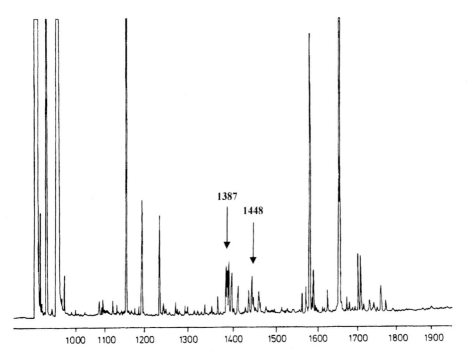

Figure 3. HRGC of a heart-cut obtained by HRGC separation of the hydrocarbon fraction isolated from hops. Compound 23a (RI: 1387) and compound 23b (RI: 1448)

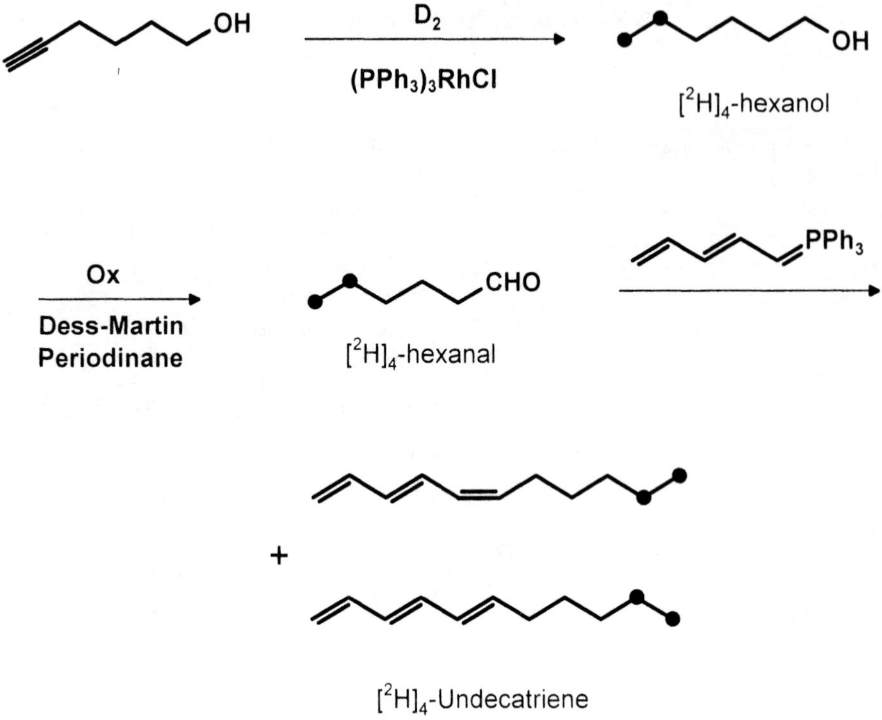

Figure 4. Synthetic route used for the preparation of [²H]₄-1,(E)3,5-undecatriene

Table IV. Concentrations[a] of selected (FD ≥ 128) key odorants in dry hop cones of the variety Spalter select

Odorant	Odor thresholds [ng/L in air]	Conc. (mg/kg dried hops)
Myrcene	n.d.	3200
Linalool	0.6	110
Nonanal	4.5	5.8
Methyl 2-methylbutanoate	0.45	0.570
1,(E)3,(Z)5-Undecatriene	0.003	0.015
Ethyl 2-methylpropanoate	0.15	0.260
trans-4,5-Epoxy-(E)-2-decenal	0.002	0.210
(Z)-1,5-Octadien-3-one	0.005	0.001
1,(E)3,(Z)5,9-Undecatetraene	0.01	0.010

[a] Except for myrcene and linalool, the concentrations were determined by stable isotope dilution assays [Steinhaus and Schieberle, J. Agric. Food Chem., 2000, in preparation].

n.d.: not determined.

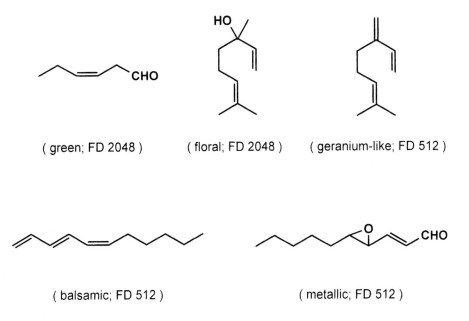

(green; FD 2048)　　　(floral; FD 2048)　　　(geranium-like; FD 512)

(balsamic; FD 512)　　　　　　　　　(metallic; FD 512)

Figure 5. Structures of the most odor-active compounds identified in fresh, undried hops

Conclusions

The data confirm earlier results of the literature on the key role of linalool and myrcene in the overall hop aroma. In addition, 4,5-epoxy-(E)-2-decenal (metallic), 1,(E)3,(Z)5-undecatriene (balsamic, fresh), the 1,(E)3,(Z)5,9-undecatetraene (balsamic) and 1,(Z)5-octadien-3-one (geranium-like) which were previously unknown as hop constituents, are suggested as further key hop aroma compounds. The drying procedure applied in hop manufacturing leads to a significant change in the overall aroma of the fresh hop cones. Based on the results presented here, in particular, the significant degradation of (Z)-3-hexenal during drying is undoubtedly responsible for the loss of the green, grassy character of the fresh hops. 4,5-Epoxy-(E)-2-decenal has previously been identified as the cause for a metallic off-odor in fats (10). Its formation was clearly demonstrated from glyceryl 13-hydroperoxy-9,11-octadecadienoate as the precursor. However, because the epoxyaldehyde was detected also in the fresh hop cones a biosynthetic formation also seems to be possible.

References

1. Chapman, A.C. *J. Chem. Soc. Trans.* **1895**, *67*, 54-63.
2. Chapman, A.C. *J. Chem. Soc. Trans.* **1895**, *67*, 780-784.
3. Semmler, B. *Ber. Dt. Chem. Ges.* **1911**, *44*, 2009-2011.
4. Nijssen, L.M.; Visscher, C.A.; Maarse, H.; Willemsens, L.C.; Boelens, M.H. In: *Volatile Compounds in Food*. TNO, Zeist, The Netherlands, **1996**, No. 62.
5. Guadagni, L.M.; Buttery, R.G.; Harris, J. *J. Sci. Food Agric.* **1966**, *17*, 142-144.
6. Schieberle, P. (1995) In: Characterization of Foods: Emerging Methods. Goankar, A. (Ed.), Elsevier, Amsterdam, **1995**, pp. 403-341.
7. Steinhaus, M.; Schieberle, P. *J.Agric Food Chem*, **2000**, in press.
8. Engel, W.; Bahr, W.; Schieberle, P. *Eur. Food Res. Technol.* **1999**, *209*, 237-241.
9. Fielder, S.; Rowan, D.D. *J. Labelled Comp. Radiopharmaceuticals* **1995**, *36*, 465-470.
10. Gassenmeier, K.; Schieberle, P. *J. Am. Oil Chem. Soc.* **1994**, 71:1315-1319

Chapter 4

Application of a Comparative Aroma Extract Dilution Analysis to Monitor Changes in Orange Juice Aroma Compounds during Processing

A. Buettner and P. Schieberle

Deutsche Forschungsanstalt fuer Lebensmittelchemie, Lichtenbergstrasse 4, D–85748 Garching, Germany

The overall flavor changes occurring during the manufacturing of processed orange juice such as juice reconstituted from concentrate were monitored by application of a comparative aroma extract dilution analysis (cAEDA). A comparison of the etheral extracts obtained from freshly squeezed and from processed juice, respectively, showed that the main differences were due to significantly higher Flavour Dilution (FD) factors of the odorants acetaldehyde (fresh, pungent) and (Z)-hex-3-enal (green) in the fresh juice. In contrast to this, the FD factors of several odor-active terpenoid compounds such as limonene, α-pinene and linalool, as well as 3-isopropyl-2-methoxypyrazine and vanillin were higher in the reconstituted juice. Additionally, the caraway-like smelling carvone was detected with a much higher FD factor in the processed juice.

Although food processing has been much improved within the last decades, flavor changes induced by, e.g. concentration steps, pasteurization and/or storage are still very common. Especially the delicate, but sensitive flavor of citrus juices is altered significantly e.g. during the process of juice concentration.

Citrus fruit flavors are known to consist of a variety of odorants eliciting very different flavor attributes, such as fresh-pungent, fruity-sweet, fatty or terpeny. However, when present in certain quantities, these odorants create the characteristic smell associated with a certain citrus fruit variety.

Recently, the key odorants in freshly squeezed orange and grapefruit juices have been characterized, and it showed up that the overall citrus-like aromas of oranges and grapefruits have many odor-active compounds in common (1, 2). However, the characteristic smell of grapefruit juice was attributed to two extremely odor-active trace compounds, the catty smelling 4-mercapto-4-methylpentan-2-one and 1-*p*-

menthene-8-thiol with a grapefruit-like odor quality. Due to their extremely low concentrations, such odorants are very difficult to monitor by use of common analytical methodologies such as GC-MS, etc. (*3*). Recently, cAEDA has been proposed as a very useful tool to objectify flavor differences between black and white roasted sesame seeds (*4*), flavor defects occurring during storage of beer (*5*) or buttermilk (*6*) and several other food materials, e.g., butter oil (*7*), extruded oat meal (*8*), trout (*9*), soya bean oil (*10*), as well as flavor changes induced during processing of foods such as white sesame seeds (*11*), strawberries (*12*) or apples (*13*).

The prerequisites for the performance of the comparative Aroma Extract Dilution Analysis (cAEDA) are as follows: the same amounts of the two food batches to be compared are used in the identical isolation procedure to obtain the volatile compounds of each sample. Both solvent extracts are then concentrated to exactly the same volume and an AEDA is subsequently performed for both samples in the same way and in parallel. Similarly, comparative aroma dilution analysis (cADA) using static headspace-olfactometry (SHO, (*14*)) can be performed to compare the highly volatile fraction of two food batches provided that exactly the same amounts of food material are used and analyses are carried out in exactly the same way.

When comparing the FD factors obtained from cAEDA one has to keep in mind that only FD-differences exceeding more than two dilution steps can be regarded as significant. As discussed previously, differences amounting to only one or two FD factors are within the error of the methodology (*15*).

Following, our results on the cAEDA of freshly hand-squeezed orange juice and processed juice reconstituted from concentrate, both made from the same orange variety (Valencia late), are presented.

Experimental Procedures

Materials

Fresh oranges (*Citrus sinensis* (L.) Osbeck; cultivar Valencia late, grown in Argentinia) were purchased from a local market and were used within two days. The freshly squeezed juice (FSJ) was obtained by careful hand-squeezing of the fruits in a kitchen juicer immediately before use. Reconstituted orange juice (ROJ) made from Valencia late oranges was obtained from a local supplier.

The isolation of the juice volatiles by solvent extraction using diethylether was performed as described previously (*2*). For the comparative aroma extract dilution analysis (cAEDA) the same amounts of juice (600 mL) and solvent (500 mL) were used. The workup procedure and the following concentration of the solvent extracts to the same volume (400 µL) were performed for both juice samples in the same way (*2*). The odorants in both extracts were then screened by AEDA by sniffing extracts of the same dilution in parallel and the odorants were identified by comparison with the reference substances on the basis of the following criteria: retention index (RI) on three stationary phases of different polarity (FFAP, SE-54; OV-1701), mass spectra obtained by MS (EI) and MS (CI), and odor quality as well as odor intensity perceived at the sniffing-port.

Comparative Aroma Extract Dilution Analysis (cAEDA)

The FD factors of the odor-active compounds in both samples were determined by cAEDA in parallel, each sniffing analysis was performed on the following dilution series: The original extract (400 µL) from 600 mL of juice was stepwise diluted with diethylether (1+1) until no odorant was detectable by sniffing of the highest dilution. HRGC-O was performed with aliquots (0.5 µL) using an FFAP capillary column. A total number of three experienced sniffers was used to perform the cAEDA experiments. Their response to the individual compounds did not differ by more than 2 FD-factors. Within two consecutive runs, the odor-active volatiles of the same dilutions of both extracts were evaluated.

High-resolution gas chromatography-olfactometry (HRGC-O) and mass spectrometry (HRGC-MS)

HRGC was performed with a Type 8000 gas chromatograph (Fisons Instruments, Mainz, Germany) using the following fused silica capillaries: free fatty acid phase (FFAP; 30 m x 0.32 mm i.d., 0.25 µm d_f; Chrompack), SE-54 (30 m x 0.32 mm i.d., 0.25 µm d_f; J&W Scientific, Fisons Instruments), and OV-1701 (30 m x 0.32 mm i.d., 0.25 µm d_f; Chrompack). Gas chromatographic conditions were the same as described previously (1). MS analysis was performed with a MS 8230 (Finnigan MAT, Bremen, Germany) in tandem with the capillaries described above. Mass spectra in the electron impact mode (MS-EI) were generated at 70 eV and in the chemical ionization mode (MS-CI) at 115 eV with isobutane as the reactant gas.

Quantification of Selected Flavor Compounds

Determination of odorants by static headspace-stable isotope dilution assays (SHA-SIDA)

Freshly squeezed or processed orange juice, respectively, was poured into saturated $CaCl_2$-solution, the vessel sealed immediately with a septum and spiked with known amounts of the labeled standards [3,3,4,4-$^2H_4$]-hexanal, [3,4-$^2H_2$]-(Z)-hex-3-enal, [3,3,4,4-$^2H_4$]-octanal, [2,2,2-$^2H_3$]-ethyl hexanoate and [2,2,2-$^2H_3$]-ethyl butanoate (16). [1,2-$^{13}C_2$]-acetaldehyde was used for the quantification of acetaldehyde. After stirring for 30 min to reach equilibration, aliquots of the headspace were withdrawn with a gastight syringe and analyzed using the HRGC-MS system Incos XL (Finnigan) fitted with the fused silica capillary Rtx-5 (30 m x 0.53 mm i.d., 1.5 µm d_f, Amchro, Sulzbach/Taunus, Germany) (17). The mass spectral data in the CI mode were obtained at 115 eV with methane as the reagent gas. At least four replicates were performed. As an example, Figure 1 shows the mass traces of ethyl butanoate and ethyl hexanoate and their labeled standards obtained during SHA-SIDA.

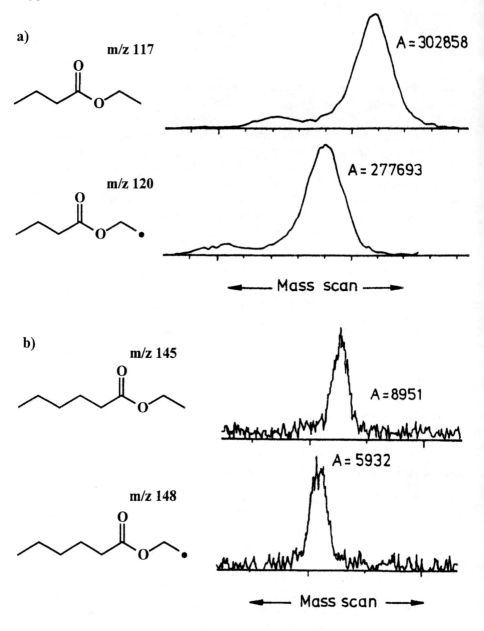

Figure 1. Mass traces of the static headspace-stable isotope dilution assays of
a) ethyl butanoate and [2,2,2-²H₃]-ethyl butanoate
b) ethyl hexanoate and [2,2,2-²H₃]-ethyl hexanoate.
(●) Position of the [²H]-atoms, (A) Intensity of the fragment ions.
25 mL of juice were used.

Results and Discussion

The sensory evaluation of the volatiles isolated from freshly hand-squeezed (FSJ) and processed orange juices (ROJ) made from Valencia late oranges by means of the comparative aroma extract dilution analysis resulted in the FD chromatograms displayed in Figure 2.

In both juices more than 28 odor-active regions with Flavor Dilution (FD) factors >32 were detected. In combination with the identification experiments (Table I) the results revealed the highest FD factors for the fruity smelling ethyl butanoate (FD 1024), the grassy (Z)-hex-3-enal (FD 512) in the FSJ, followed by the fruity-sweet smelling winelactone (3a,4,5,7a-tetrahydro-3,6-dimethyl-2(3H)-benzofuranone; FD 256), the fruity ethyl 2-methylbutanoate (FD 128) and tr-4,5-epoxy-(E)-dec-2-enal with metallic odor (FD 128). Except (Z)-hex-3-enal, all these odorants were also found with quite high FD factors in the ROJ. In addition, tr-4,5-epoxy-(E)-dec-2-enal, oct-1-en-3-one and ethyl 3-hydroxyhexanoate were significantly reduced in the processed juice sample.

In contrast, the caraway-like smelling carvone was detectable as the only "new" odorant in the processed juice. However, the intensities of several other odorants were increased in the ROJ such as α-pinene, limonene, octanal and linalool. Other odorants exhibiting higher FD-factors in the processed juice sample were ethyl 2-methylbutanoate, 3-isopropyl-2-methoxypyrazine, 2/3-methylbutanoic acid, (E,E)-nona-2,4-dienal and vanillin (Table 1).

The FD factors of the remaining eleven odorants detected in both samples did not differ significantly.

The observed flavor differences between the freshly squeezed orange juice and the processed juice made from concentrate can be explained by chemical and by physical processes. For instance, linalool, limonene and octanal are typical constituents of orange peel oil. Their enhanced FD-factors (or amounts, respectively) in the reconstituted juice indicate that the higher pressure during industrial juice production leads to their increased release from the peel into the juice. Furthermore, it is common industrial practice to enhance the fresh, citrus-like note of processed orange juices by adding distinct amounts of peel oil to the juice (18). Additionally, linalool may be liberated from its glucoside during heating of the juice. An increase of this odorant in pasteurized orange juice was also observed by Marin et al. (19) using the CHARM-technique for the detection of flavor changes during processing. Comparatively higher amounts of the peel-oil odorants octanal, decanal and linalool in juice made from concentrate were also reported previously by Nisperos-Carriedo and Shaw (20).

In addition, the results of the cAEDA indicate that the fresh, grassy smelling (Z)-hex-3-enal is significantly reduced in the reconstituted juice. The reduction of this important aldehyde of the fresh juice has not been reported previously, but its degradation is likely to cause a major aroma difference in the processed juice. An explanation for its absence in ROJ is the lability of (Z)-hex-3-enal which is easily degraded in acidic aqueous solutions. Some possible breakdown pathways are given

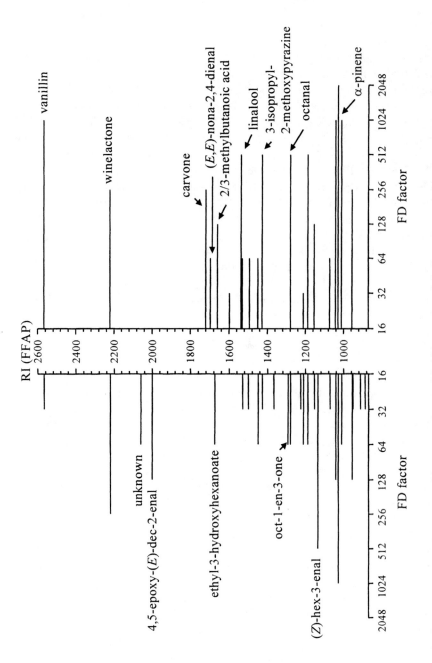

Figure 2. *Comparative aroma extract dilution analysis of freshly squeezed (left) and processed, reconstituted juice (right) made from Valencia late oranges.*

Table I. FD factors (FD ≥ 32 in at least one sample) of important odorants in freshly hand-squeezed juice (FSJ) and reconstituted juice (ROJ) made from Valencia late-oranges.

Odorant	FD factor [a]		RI on FFAP
	FSJ	ROJ	
2/3-Methylbutanal	32	nd	913
Ethyl 2-methylpropanoate	128	256	955
α-Pinene	64	1024	1010
Ethyl butanoate	1024	2048	1028
Ethyl 2-methylbutanoate	128	1024	1041
Hexanal	32	64	1072
(Z)-Hex-3-enal	512	nd	1135
Myrcene	32	128	1154
Limonene	64	512	1188
2/3-Methylbutanol	64	32	1211
Octanal	64	512	1279
Oct-1-en-3-one	64	8	1292
3-Isopropyl-2-methoxypyrazine	32	512	1427
Methional	64	64	1449
Acetic acid	16	32	1450
Decanal	16	64	1493
(Z)-Non-2-enal	32	nd	1500
(E)-Non-2-enal	32	64	1530
Linalool	16	512	1537
1-p-Menthene-8-thiol	8	32	1598
2/3-Methylbutanoic acid	4	128	1660
Ethyl 3-hydroxyhexanoate	64	8	1674
(E,E)-Nona-2,4-dienal	4	64	1696
Carvone	nd	256	1720
tr-4,5-Epoxy-(E)-dec-2-enal	128	16	2000
Unknown	64	nd	2059
3a,4,5,7a-Tetrahydro-3,6-dimethyl-2(3H)-benzofuranone	256	256	2220
Vanillin	32	1024	2567

a) The Flavor Dilution (FD) factor was determined in etheral extracts containing the juice volatiles. Analyses were performed by two assessors in duplicates.

b) nd: The compound was not detected in one of the concentration steps of the sample applied for cAEDA.

in Figure 3, involving e.g. aldol-, retroaldol-, addition- and oxidation-reactions. Similar pathways can be proposed for the degradation of other labile compounds like *tr*-4,5-epoxy-(*E*)-dec-2-enal and (*Z*)-non-2-enal which were also found in lower concentrations in the processed juice.

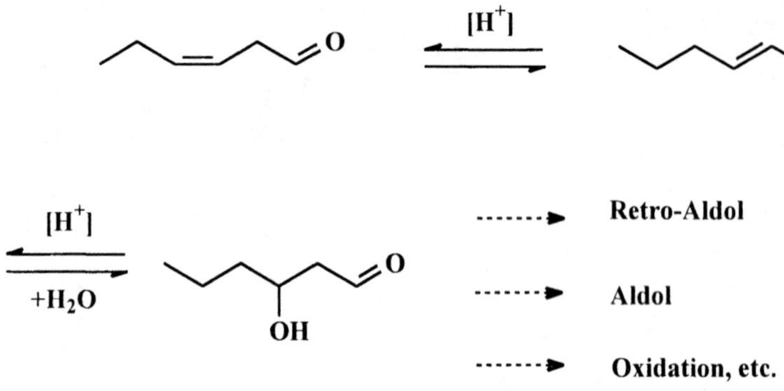

Figure 3. Possible degradation pathways of (Z)-hex-3-enal.

In previous investigations, the formation of carvone was already suggested to cause off-flavor in orange juice (*21, 22*). A formation pathway for this caraway-like smelling compound via the corresponding hydroperoxide (Figure 4) was already proposed by Farmer and Alvapillai (*23*). Although the stereochemical composition of carvone was not determined in this investigation, the (*S*)-enantiomer seemed to be indeed the predominant odor contributor because a caraway-like odor quality was perceived during cAEDA of processed orange juice. In contrast, (*R*)-carvone is characterized by a minty smell.

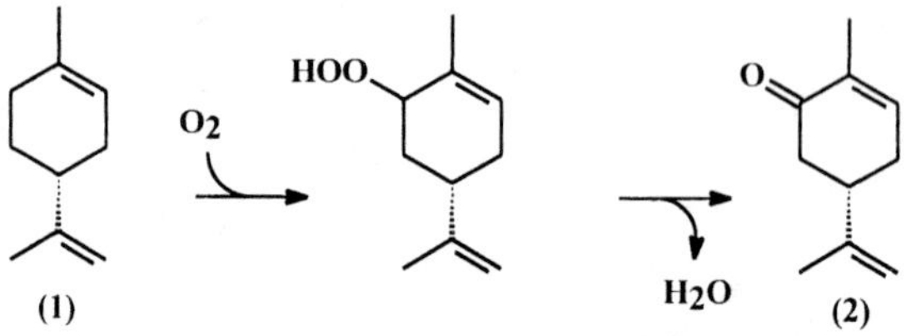

Figure 4. Formation of (S)-carvone (2) from (R)-limonene (1) according to Farmer and Alvapillai (1942).

The increased amounts of vanillin in the ROJ can be explained by its formation from ferulic acid as discussed by Peleg et al. (*24*). The acid-catalyzed formation pathway proposed by the authors is displayed in Figure 5.

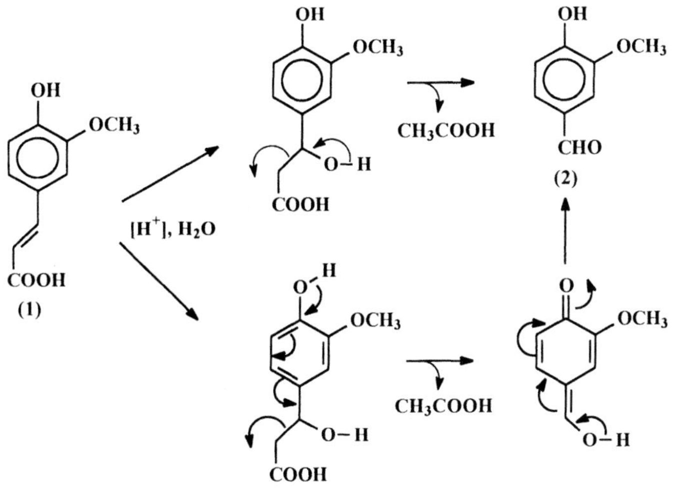

Figure 5. Acid-catalysed formation of vanillin (2) from ferulic acid (1) according to Peleg et al. (1992).

Previous investigations have shown (16) that, apart from the odorants discussed above, acetaldehyde is another very important contributor to the fresh, pungent odor quality of freshly hand-squeezed orange juices. For that reason the concentrations of acetaldehyde were determined in both juices by static headspace-stable isotope dilution analysis. Additionally, the aldehyde was quantified in a commercial, non-reconstituted but pasteurized juice. The results given in Figure 6 clearly indicate that significant amounts of acetaldehyde are lost during the concentration of orange juice undoubtedly due to the high volatility of the aldehyde but probably also due to chemical reactions like oxidation, condensation, etc..

In comparison to the fresh juice the amounts of acetaldehyde were reduced by a factor of about 40 in the juice made from concentrate while in the pasteurized juice about half of the original concentration was still present. In general, the flavor of the pasteurized juice was described as more fresh and pleasant than juice made from concetrate and was said to be more similar to freshly hand-squeezed orange juice (data not shown).

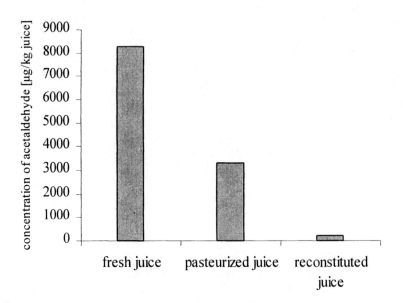

Figure 6. Concentrations of acetaldehyde in freshly hand-squeezd orange juice and in two commercial juices either reconstituted from concentrate or only pasteurized, respectively.

Changes in selected odorants occurring in the fruits

When comparing fresh and processed juices, variations of the concentrations of the odorants, e.g., due to the maturation of the fruits have to be considered. As shown in Figure 7, esters and aldehydes were shown to decrease significantly in orange fruits during a storage period of three weeks (4 °C, relative atmospheric moisture 80 %). A comparison of the quantitative data determined before and after storage of the fruits by use of static headspace-stable isotope dilution assays showed that mainly hexanal and (Z)-hex-3-enal were significantly reduced while the esters still amounted half of the original concentrations. Octanal did not decrease significantly.

However, a calculation of the odor activity values before and after storage of the fruits revealed that (Z)-hex-3-enal is still one of the most potent odorants even after storage (Table II).

In contrast to this, the differences observed for (Z)-hex-3-enal between the fresh orange juice and the processed juice made from concentrate were by far higher than those observed for the stored oranges, showing a concentration difference of a factor of about 500 (as related to the FD factor).

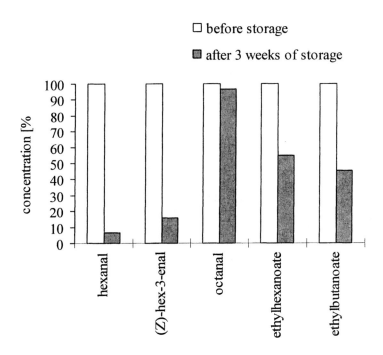

Figure 7. Relative concentration changes of selected odorants in Valencia late-oranges as determined after three weeks of storage of the intact fruits (initial concentrations of the odorants correspond to 100 %).

Table II. Comparison of the odor activity values of selected odorants in fresh
Valencia late-oranges and after three weeks of storage.

odorant	odor activity value [a]	
	before storage	after 3 weeks of storage
hexanal	19	1
(Z)-hex-3-enal	747	119
octanal	7	7
ethyl hexanoate	13	7
ethyl butanoate	1192	543

a)   The odor activity values were calculated by dividing the concentration of the odorants
determined in the food sample by its orthonasal odor threshold in water. The quantitative data
are mean values of duplicates.

Conclusions

Application of cAEDA on two solvent extracts obtained from freshly, hand-squeezed
orange juice and reconstituted orange juice revealed significant flavor differences
between the two samples. These differences could be mainly related to an increase of
carvone in the processed orange juice and a reduction of the characteristic fresh-
pungent odorant acetaldehyde and the green smelling (Z)-hex-3-enal. The results
clearly indicate that cAEDA is a valid method for the detection of flavor differences
between food samples. Using this technique it is even possible to detect flavor
changes which are caused by extremely potent odorants occurring in only trace
amounts without the need to perform time-consuming quantitative analyses.

References

1   Buettner, A.; Schieberle, P. *J. Agric. Food Chem.* **1999**, *47*, 5189-5193.
2   Hinterholzer, A.; Schieberle, P. *Flavour Fragr. J.* **1998**, *13*, 49-55.
3   Buettner, A. *Wichtige Aromastoffe in frisch gepressten Citrusfruchtsaeften aus
     verschiedenen Orangenvarietäten (Citrus sinensis (L.) Osbeck) sowie Grapefruit
     (Citrus paradisi Macf.)*; Herbert Utz Verlag Wiss.: Munich, Germany, 1999.
4   Schieberle, P. In: *Olfaction and Taste*-ISOT XI; Kurihara, K.; Suzuki, N.;
     Ogawa, H.; eds; Springer, Tokyo, 1994.
5   Schieberle, P. *Z. Lebensm. Unters. Forsch.* **1991**, *193*, 558-565.
6   Heiler, C.; Schieberle, P. *Lebensm. Wiss. Technol.* **1996**, *29*, 460-464.
7   Widder, S.; Sen, A.; Grosch, W. *Z. Lebensm. Unters. Forsch.* **1991**, *193*, 32-35.
8   Guth, H.; Grosch, W. *Z. Lebensm. Unters. Forsch.* **1993**, *196*, 22-28.
9   Grosch, W.; Milo, C.; Widder, S. In: *Trends in Flavour Research*; Maarse, H.;
     Van der Heij, D.G.; eds; Elsevier, Amsterdam, 1994.

10   Guth, H.; Grosch, W. *Lebensm. Wiss. Technol.* **1990**, *23*, 59-65.

11   Schieberle, P. In: *Progress in Flavour Precursor Studies*; Schreier, P.; Winterhalter, P.; eds; Allured Publishing Corporation, Carol Stream, USA, 1993.

12   Schieberle, P. In: *Trends in Flavour Research*; Maarse, H.; Van der Heij, D.G.; eds; Elsevier, Amsterdam, 1994.

13   Zhou, P.G.; Cox, J.A.; Roberts, D.D.; Acree, T.E. In: *Progress in Flavour Precursor Studies*; Schreier, P.; Winterhalter, P.; eds; Allured Publishing Corporation, Carol Stream, IL, USA, 1993.

14   Holscher, W.; Steinhart, H. *Z. Lebensm. Unters. Forsch.* **1992**, *195*, 33-38.

15   Schieberle, P. In: *Characterization of food: emerging methods*; Gaonkar, A.G.; ed.; Elsevier Science B.V.; 1995.

16   Buettner, A.; Schieberle, P. *J. Agric. Food Chem.* **2000**, in preparation.

17.   Guth, H.; Grosch, W. *J. Agric. Food Chem.* **1994**, *42*, 2862-2866.

18   Nagy, S.; Shaw, P.E. In: *Food flavours part C, the flavour of fruits*; Morton, I.D.; Mac Leod, A.J.; eds; Elsevier, New York, 1990.

19   Marin, A.B.; Acree, T.E.; Hotchkiss, J.H.; Nagy, S. *J. Agric. Food Chem.* **1992**, *40*, 650-654.

20   Nisperos-Carriedo, M.; Shaw, P.E. *J. Agric. Food Chem.* **1990**, *38*, 1048-1052.

21   Proctor, B.E.; Kenyon, E.M. *Food Technol.* **1949**, *3*, 387-392.

22   Beard, J.H.; Fletcher, B.C., Van Berestyn, E.C.H. *J. Sci. Food Agric.* **1972**, *23*, 207-213.

23   Farmer, E.H.; Alvapillai, S. *J. Chem. Soc.* **1942**, 121-124.

24   Peleg, H.; Naim, M.; Zehavi, U., Rouseff, R.L.; Nagy, S. *J. Agric. Food Chem.* **1992**, *40*, 764-767.

Chapter 5

Trace Compound Analysis for Off-Flavor Characterization of Micromilled Milk Powder

Martin Preininger and Frank Ullrich

Kraft Jacobs Suchard R&D, Unterbiberger Strasse 15, D–81737 Munich, Germany

Abstract

Aroma compounds causing the off-flavor in micromilled and stored milk powder (MMP) were detected and identified by comparative GC/Olfactometry and GC/MS of aroma distillates from untreated standard milk powder (MP) and MMP. In addition to aldehydes and ketones from lipid peroxidation, *2-aminoacetophenone* (AAP) was detected only in MMP and was present at high aroma intensity (musty, wet popcorn odor). AAP was not found in MP.

AAP was quantified by GC/MS using internal surrogate standards, and occurred in MMP at high Odor Activity Value (OAV = concentration/odor threshold), indicating its strong contribution to the off-flavor. A method for convenient synthesis of deuterium labeled d_3-AAP was developed for use in precise GC/MS quantitation (Stable Isotope Dilution Assay). However, the application of d_3-AAP for quantitation of AAP in MP failed because of matrix-effected deuterium-loss.

Introduction

After micromilling sweet-dairy smelling MP by high speed collision, a musty off-flavor developed within days of storage. The off-flavor was reminiscent of wet-dog fur odor. The objective of this study was to identify potent aroma compounds causing the off-flavor in MMP by GC/Olfactometry and GC/MS of aroma distillates. In addition, methods should be evaluated for quantitation of a key off-flavor compound in order to analytically measure the off-flavor development.

Material and Methods

Milk Powder Samples

Standard milk powder (MP)
MP consisted of 75 % (w/w) skim milk powder and 25 % whey powder. 90 % of the particles had 360 µm particle size.

Micromilled milk powder (MMP)
MP was micromilled to particle size distribution 90 % ≤ 7.7 μm and 50 % ≤ 3.6 μm by high speed collision under controlled atmosphere (88 % N_2, 12 % O_2) and stored at 16 °C for 10 months (qualitative analysis) and from one day up to 12 months (quantitative analysis).

In the 12 months stored sample, the moisture was 3.44 % (w/w) and the total fat content was 0.9 % at the time of quantitative analysis. MP and MMP were stored with silica gel pouches at 16°C in odorless plastic bags in the dark according to pilot plant conditions.

Sample Preparation

Chemicals
All chemicals were purchased from Aldrich (Steinheim, Germany) at the highest purity available, except for deuterated d_3-2-aminoacetophenone, which was synthesized in this study.

Solvent Extraction
For qualitative analysis, milk powder (100 g) was power stirred in freshly distilled diethylether (200 mL) in the dark under argon at room temperature for 30 min. Deionized, sonicated water (150 g) was added and stirring continued for an additional 50 min.

Solvent Assisted Flavor Evaporation (SAFE®)
The slurry from solvent extraction was distilled by SAFE® (1) using high vacuum (ca. 5 x 10⁻⁵ mbar) for 45 min at 40 °C during sample addition under argon. After sample addition, distillation was continued for an additional 75 min at 50 °C. The distillate was washed with saturated aqueous sodium chloride (40ml), and the etherous phase was dried over sodium sulfate (anhydrous). The dried etherous phase was then concentrated to 2 mL via Vigreux column distillation at 40°C before GC/O and GC/MS analysis.

For quantitative analysis, 50 g milk powder, 100ml diethyl ether, 75g water and 20ml sodium chloride solution were used, respectively. Internal standard compounds (3,4-dimethoxyacetophenone, 4-decanone, d_3-2-aminoacetophenone; see Results) were added to the diethylether during the solvent extraction step. The SAFE® distillate was concentrated to 300 μL before GC/MS analysis. Other parameters in the sample preparation procedure were kept as described above for qualitative analysis.

GC/O Analysis
The SAFE® distillate (1 μL) was injected into a HRGC Mega Series 5300 (Carlo Erba, Milan Italy) equipped with an on-column injector (Carlo Erba). The sample was separated by a CP-FFAP fused silica capillary (25 m, 0.32 mm i.d., 0.3 μm film thickness; Chrompack Inc.) with the following oven and injector program: The initial temperature of 35 °C was held for 2 min and then ramped to 60 °C at a rate of 40 °C/min. The temperature of 60 °C was held for 2 min and then ramped to 240 °C

at a rate of 6 °C/min. The final temperature of 240 °C was held for 15 min. The capillary effluent was split (1:1) to a flame ionization detector and a sniffing port held at 200 °C.

GC/MS Analysis
The SAFE® distillate (1 µL) was injected into a HRGC Mega Series 5300 (Carlo Erba, Milan Italy) equipped with a KAS 3 cold on-column injector (Gerstel, Munich, Germany). The sample was separated by a HP-FFAP crosslinked fused silica capillary (50 m x 0.32 mm i.d. x 0.5 µm film thickness, Agilent, Taufkirchen, Germany) with the following oven program: The initial temperature of 35 °C was ramped to 230 °C at a rate of 6 °C/min. The final temperature of 230 °C was held for 15 min. The sample was analyzed using a mass selective detector, MSD 5973 (Agilent). The MSD 5973 was operated either in EI-SCAN mode (35-250 amu) or in SIM mode as specified. Data analysis was performed using the HP-Chemstation of the MSD 5973. For the different lots of MMP analyzed, the corresponding MP was also analyzed before micromilling.

Synthesis of deuterated 2-aminoacetophenone (d_3-AAP)
AAP (7.2 mg) was dissolved in deuterium oxide (5.0 mL), adjusted to pH 14.0 with sodium deuteroxide (40 % in D_2O) and vigorously stirred at room temperature for 16 h. After deuteration the mixture was extracted with diethylether (3 x 7 mL) using a 10 mL Mixxor® (Aldrich) and dried over sodium sulfate (anhydrous). The concentration of the etherous d_3-AAP solution was determined by GC/MSD 5973 SIM with AAP as internal standard and assuming a response factor of 1.0.

d3-AAP Stability Control
Water (HPLC grade, 1.0 mL) was added to a solution of *d3-AAP* (5.13 µg) in ether (4 mL) and adjusted to pH 6.27 with hydrochloric acid (0.1 N). The mixture was vigorously stirred at room temperature for 80 min. The etherous phase was then separated, washed with saturated aqueous sodium chloride (1.5 mL) by the Mixxor®, and dried with sodium sulfate (anhydrous) before GC/MS, similar to the procedure of AAP quantitation from milk powder.

Results and Discussion

Identification of Potent Aroma Compounds in MP and MMP
The aroma compounds from MP and MMP were isolated by high vacuum distillation of a solvent extract. The distillate of MMP represented the off-flavor well, possessing an intense musty, stale odor. The MP distillate had a weak, sweet odor like caramelized milk. The distillates were analyzed by comparative GC/O and GC/MS for detection and identification of potent aroma compounds which may cause the flavor difference between the samples.
2-Aminoacetophenone (AAP) was detected and identified only in the MMP distillate by an intense musty, wet-popcorn odor and by its retention index in GC/O (Table 1, No.1). Based on the GC/O results, the characteristic mass traces of AAP (m/z 135,

120 and 92) were extracted from the total ion chromatograms after GC/MS-SCAN analysis of the MP and MMP distillates (Figure 1 and 2). Peaks of AAP mass traces occurred only in the mass chromatogram of the MMP distillate. The mass traces of AAP were overlayed by a peak of m/z 243 which originated from an unknown compound. This inhibited direct identification of AAP by MS library search (Figure 3). After background substraction the identity of AAP was subsequently confirmed by its mass spectrum (Figure 4). AAP was not found in MP by GC/O and by GC/MS-SCAN (Figure 1) and GC/MS-SIM (data not shown).

Metallic, green, fatty smelling aldehydes and ketones, which derive from lipid-peroxidation of milk-fat, were detected by GC/O either only in MMP, or at much higher aroma intensity in MMP, compared to MP (Table 1, No.6-22).

Alkanals were found by GC/MS at much higher level in MMP than in MP, using acetic and octanoic acid as naturally occurring internal reference compounds (Figure 5 and 6).

The aroma intensity of sweet caramelized smelling Furaneol® was found by GC/O to be much lower in MMP compared to MP (Table 1, No.23).

In summary GC/O data were interpreted in the following way. In MMP, the sweet, caramelized smelling compounds (Table 1, No.23-27) may be lower in concentration and may be masked by increased concentrations of musty, green, metallic smelling off-flavor compounds, compared to MP. The off-flavor compounds may have derived from lipid and amino acid degradation (2, 3, 4) during micromilling and storage of milk powder.

Quantitation of 2-Aminoacetophenone (AAP) via Surrogate Standards

In order to measure the off-flavor intensity, AAP was quantified by GC/MS-SIM from MMP using the internal surrogate standards, 3,4-dimethoxyacetophenone (490 µg/kg) and 4-decanone (510 µg/kg). MS response factors for each standard were determined by GC/MS-SIM. Recovery factors were determined by applying the quantitative GC/MS-SIM analysis to MP which had been spiked with AAP (396 µg/kg), and which did not contain detectable amounts of AAP naturally. The recovery factors were calculated as quotients of the actually added amount of AAP and the amount of AAP measured via the internal surrogate standards, employing the MS response factors.

The recovery factors (Table 2) for AAP determined via 3,4-dimethoxyacetophenone and via 4-decanone varied greatly by the factor of 151, showing the importance of determining the recovery factors in order to obtain correct analyte concentration data of AAP.

AAP was found in 12 months stored MMP at a level of ca. 13 µg/kg (Table 3). In different lots of MMP the AAP concentrations were 9.3 µg/kg after 10 months storage and 6 µg/kg after only one day of storage. No AAP was detectable in MP from which each of the different MMP lots were produced.

AAP is a highly potent off-flavor compound in MMP due to its musty smell and high Odor Activity Value.

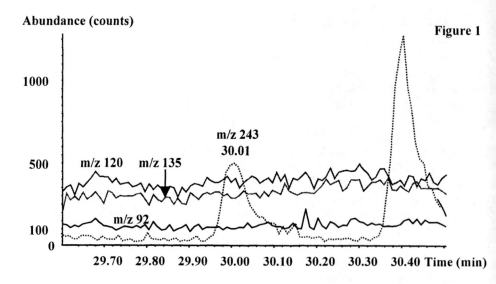

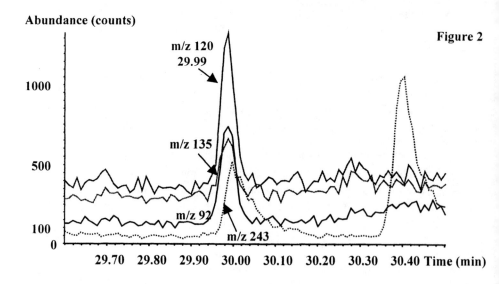

Figure 1 and 2:
Extracted ion chromatogram (GC/MS-SCAN) of distillate from standard milk powder (MP, Figure 1) and micro-milled and stored milk powder (MMP, Figure 2). 2-Amino-acetophenone (AAP) detected by its mass traces (m/z 135, 120 and 92) only in MMP. Mass trace m/z 243 stems from unknown overlaying compound.

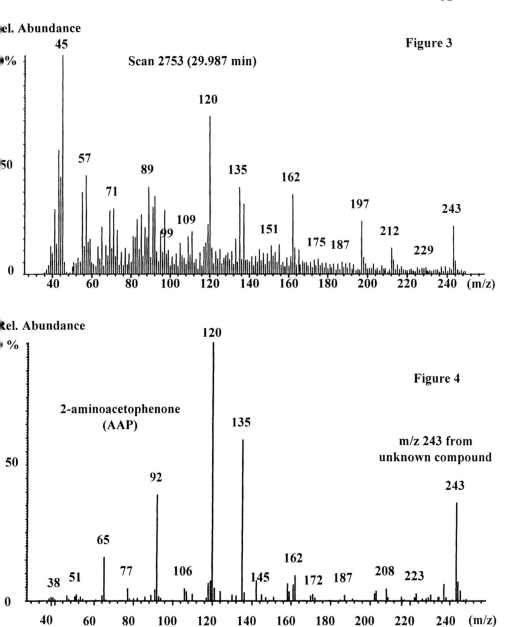

Figure 3 and 4:
EI-MS of 2-aminoacetophenone (AAP) detected in distillate from micromilled and stored milk powder without (Figure 3) and with (Figure 4) background substraction from GC/O information.

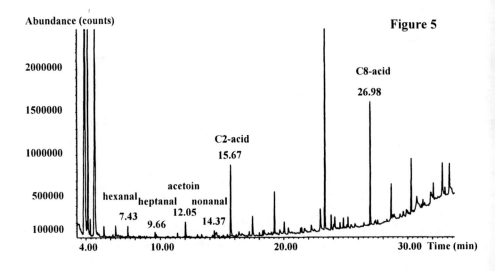

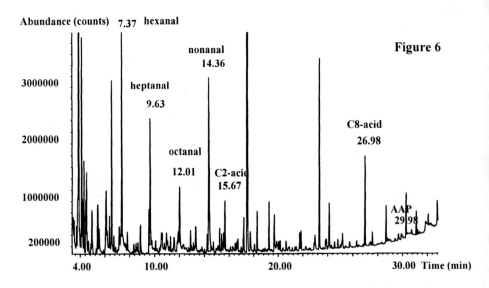

Figure 5 and 6:
GC/MS-SCAN total ion chromatogram of distillates from standard milk powder (MP, Figure 5) and micromilled, stored milk powder (MMP, Figure 6).

Table 1:

Potent aroma compounds detected by GC/O in distillates of standard milk powder (MP) and micromilled, stored milk powder (MMP) with off-flavor. Selection from 46 detected compounds showing odor intensity differences between both samples. (++++ is very high, +++ is high, ++ is medium, + is weak odor intensity, - is not detected in GC/O)

No.	RI $_{FFAP}$	odor	compound [1]	odor intensity MP	MMP
1	2227	musty, wet popcorn, tortilla	2-amino-aceto-phenone (AAP)	-	+++
2	2206	seasoning like	Sotolon	-	++
3	924	fruity, malty	3-methylbutanal	-	++
4	2068	phenolic, medicinal	4-methylphenol	-	+
5	2500	fecal, skatol	skatol	-	+
6	1244	putty	(Z)-4-heptenal	-	+++
7	1584	cucumber	(E,Z)-2,6-nonadienal	-	+++
8	2007	metallic	trans-4,5-epoxy-(E)-2-decenal	++	+++
9	1902	metallic	trans-4,5-epoxy-(E)-2-nonenal	-	++
10	1188	grassy	heptanal	-	++
11	1289	lemon, green	octanal	-	++
12	1393	green	nonanal	-	++
13	1305	mushroom	1-octen-3-one	-	++
14	1377	geranium-like	(Z)-1,5-octadien-3-one	-	++
15	1871	tallowy	12-methyltridecanal	-	++
16	1533	musty, green	(E)-2-nonenal	+	++
17	1701	fatty, deep fried	(E,E)-2,4-nonadienal	+	++
18	1811	fatty, deep fried	(E,E)-2,4-decadienal	+	+
19	1107	fruity solvent	hexanal	-	+
20	1405	mushroom	1-nonen-3-one	-	+
21	1429	fatty	(E)-2-octenal	-	+
22	1505	geranium-like	3,5-octadien-2-one	-	+
23	2037	caramel	Furaneol®	++++	++
24	1337	toasted	2-acetyl-1-pyrroline	++	++
25	1919	coconut	γ-octalactone	-	++
26	2196	coconut	δ-decalactone	+	+
27	2569	vanilla	vanillin	-	+

[1] Compounds identified by their characteristic odor, retention index (RI, n-alkanes), and GC/MS (except No. 2, 6-9, 13-15, 20, 23, 24), matching data of standard compounds.

Table 2:

MSD 5973 (SIM) response data for analyte 2-aminoacetophenone (AAP), and surrogate standards, 3,4-dimethoxyacetophenone (34DMoAP) and 4-decanone (4C10on). Recovery factors.

	RT (min)	(m/z) target ion	(m/z) qualifier ion	RespF [1]	RecF [2]
AAP via 34DMoAP	29.92	135	120	2.358	0.041
34DMoAP	33.75	165	180	---	---
AAP via 4C10on	29.92	135	120	0.474	6.202
4C10on	15.07	113	71	---	---

[1] MS Response Factor determined in duplicates at m/z peak area size of analyte in samples
[2] Recovery factors determined by GC/MS analysis of standard milk powder spiked with surrogate standards and AAP

As long as recovery factors were applied for each surrogate standard, the calculation of the AAP concentration resulted in approximately the same value, regardless of which standard was used for quantitation (Table 3).

2-amino-acetophenone (AAP) in micromilled and stored (12 months) milk powder (MMP). AAP quantified via surrogate standards, 3,4-dimethoxyacetophenone (34DMoAP) and 4-decanone (4C10on). OAV calculated via two odor threshold data.

No.	AAP analysis	concentration in MMP (μg/kg)	odor threshold (μg/kg)	OAV [3]
1	via 3,4DMoAP	12.6	0.01 [1]	1260
2	via 3,4DMoAP	12.6	0.5 [2]	25.2
3	via 4C10on	13.6	0.01 [1]	1360
4	via 4C10on	13.6	0.5 [2]	27.2

[1] *Rapp*, 1998 (5) [2] *Arnold* and *Lindsay*, 1969 (6)
[3] Odor Activity Value; OAV = concentration / odor threshold

Quantitation of 2-Aminoacetophenone (AAP) via d3-AAP

Synthesis of d3-AAP
An attempt was made to quantify AAP by GC/MS-SIM using its deuterium labeled analogue, d_3-AAP, as internal standard in a Stable Isotope Dilution Assay (SIDA). In SIDA, analyte and standard are recovered to almost the same degree during sample preparation because of their analogous chemical structure and physico-chemical properties. Therefore, SIDA does not require the determination of recovery factors (7, 8).
Dollmann et al. (9) determined AAP in wines by SIDA, and synthesized the internal standard, acetyl-group deuterated d_3-AAP, via Grignard reaction. In this present study, d_3-AAP synthesis was greatly facilitated by H/D-exchange of AAP in alkaline D_2O (Figure 7).
AAP was completely deuterated to d_2- and d_3-AAP at pH 14.0 (*Table 4 and 5, Figure 8*). In preliminary experiments, AAP was completely deuterated to d_1-, d_2-, and d_3-AAP at pH 13.7 (3 h in D_2O, alkalized with 0.1 N NaOH), but the amount of d_2- and d_1-AAP was much higher than in the pH 14 experiment (data not shown). No deuteration of AAP occurred at pH 12.9 (3 h in D_2O, alkalized with NaOH$_{solid}$; data not shown).

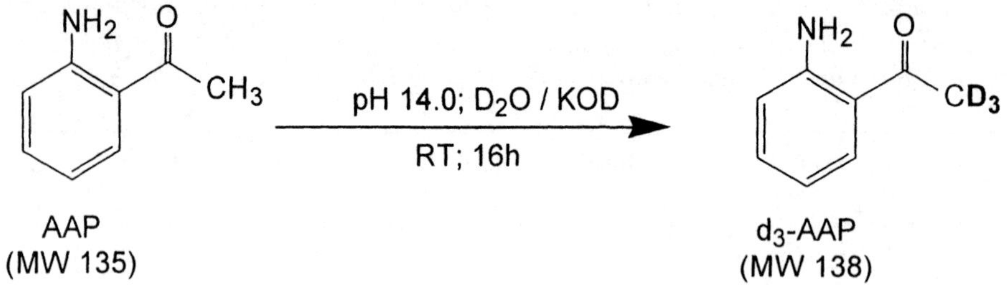

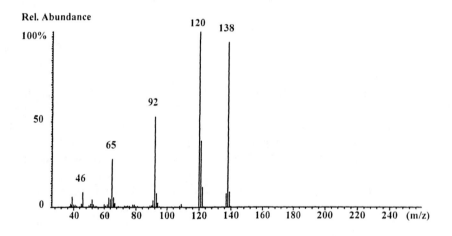

Figure 7:
Pathway for deuteration of 2-aminoacetophenone in alkaline D_2O

Figure 8:
EI-MS of d-2-aminoacetophenone (d_3-AAP) after deuteration in D_2O at pH 14.0

Table 4:
Isotope ratio of 2-aminoacetophenone (AAP) in peak area (PA) %

No.	compound	m/z	ret. time (min)	% of total PA (m/z)	PA (m/z) in % of PA (m/z 135)
1	AAP	134	n.d.	0.00	0.00
2	AAP	135	29.91	87.44	100.00
3	AAP	136	29.90	11.48	13.13
4	AAP	137	29.91	1.08	1.24
5	AAP	138	n.d.	0.00	0.00
6	AAP	139	n.d.	0.00	0.00
7	AAP	140	n.d.	0.00	0.00
8	AAP	141	n.d.	0.00	0.00

n.d. = not detected

Table 5:
Isotope ratio of d-2-aminoacetophenone (d-AAP) after deuteration in D_2O at pH 14.0

No.	compound	m/z	ret. time (min)	% of total PA (m/z)	PA (m/z x) in % of PA (m/z 138)
1	AAP	135	n.d.	0.00	0.00
2	d1-AAP	136	29.93	0.81	1.05
3	d2-AAP	137	29.92	13.65	17.53
4	d3-AAP	138	29.91	77.89	100.00
5	dn-AAP	139	29.91	7.16	9.19
6	dn-AAP	140	29.91	0.48	0.62
7	dn-AAP	141	n.d.	0.00	0.00

n.d. = not detected

d3-AAP Stability Control
The stability of d_3-AAP against reverse D/H-exchange was checked in a water/ether mixture at the pH of MMP (pH 6.29) in order to exclude AAP formation from d_3-AAP during sample preparation in quantitation of AAP in milk powder. No AAP (m/z 135) was formed from d_3-AAP (m/z 138) in this aqueous model system at pH 6.27 (Figure 9).

58

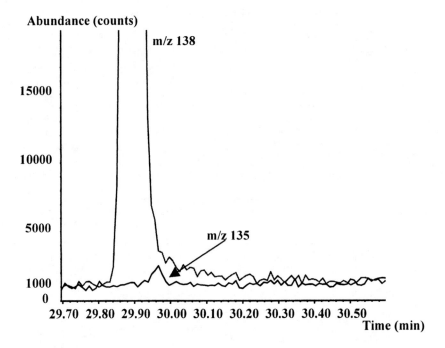

Figure 9:
Extracted ion chromatogram of d₃-AAP after treatment at pH 6.27.

Quantitation of 2-Aminoacetophenone (AAP) in Standard Milk Powder via d₃-AAP

A concentration value for AAP in clean smelling MP was needed as a reference in order to compare it to AAP concentrations in MMP which had an off-flavor. Therefore, an attempt was made to quantify AAP in MP by GC/MS using the internal standard, d_3-AAP (8.55 μg/50 g MP). AAP (m/z 135) was found at 11.5 % of the peak intensity of d_3-AAP (m/z 138; Figure 10), resulting in a calculated AAP concentration of 19.7 μg/kg in MP. Since off-flavored MMP contained only ca. 13 μg/kg (Table 3), and since no AAP was detected by GC/MS-SCAN (Figure 1) and GC/MS-SIM (data not shown) in MP which had not been spiked with d_3-AAP, the AAP found in this SIDA experiment is believed to have been formed by degradation of d_3-AAP.

AAP was obviously formed by deuterium/hydrogen exchange of d_3-AAP in the milk powder matrix, but not in the aqueous model system (Figure 9). This indicates reversible binding of AAP to the milk powder matrix, catalyzing the D/H-exchange.

Conclusions

Comparative GC/O is a suitable tool to reveal potent aroma compounds which can be used as indicator compounds in GC/MS to objectively measure the difference in flavor quality of food samples. GC/O was the key to identify 2-aminoacetophenone (AAP) as a potent off-flavor compound in micromilled milk powder (MMP).

AAP can be quantified in MMP using surrogate standard compounds, however, recovery factors must be determined and applied to obtain accurate, absolute data.

Stable Isotope Dilution Assay (SIDA) may be useful to quantify trace aroma compounds in order to avoid determination of recovery factors. In this study, acetyl-group deuterated d_3-AAP was not a suitable standard compound for SIDA because it was found to generate the analyte, AAP, during sample preparation in the food matrix. However, d_3-AAP was stable in an aqueous model system at same pH. Therefore, SIDA standards labeled in CH-acidic positions should be checked for stability in the actual sample matrix, not just in a model system. For AAP quantitation in milk powder via SIDA, a deuterated AAP standard must be labeled in non-CH-acidic position.

Acknowledgement
Thanks to A. Nowak and A. Savarino who supported this study by sample preparation. Kraft Jacobs Suchard R&D Munich provided the analytical equipment and supplies.

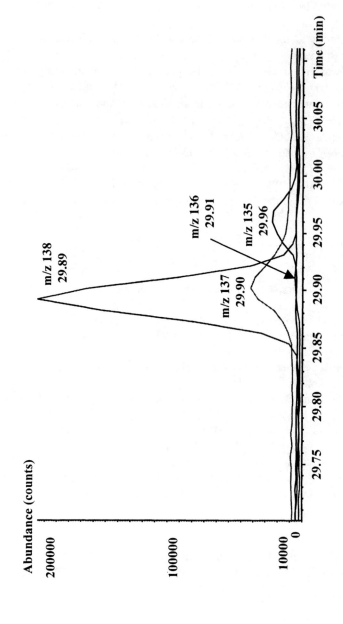

Figure 10:
Extracted ion chromatogram of d₃-AAP after spiking into, and distillation from standard milk powder (MP) at pH 6.29

Literature References

(1) Engel, W.; Bahr, W.; Schieberle, P. *Zeitschr. Lebensm. Unters. Forsch.* **1999,** *209,* 237-241.

(2) Walker, N.J.; Manning, D.J. *New Zealand J. Dairy Sci. Technol.* **1976,** *11,* 1-9.

(3) Moreaux V.; Birlouez-Aragon I. *J. Agric. Food Chem.* **1997,** *45,* 1905-1910.

(4) Dollmann B.; Schmitt A.; Köhler H.; Schreier P. *Vitic. Enol. Sci.* **1996,** *51 (2),* 122-125.

(5) Rapp A. *Nahrung* **1998,** *42 (6),* 351-363.

(6) Arnold R.G.; Lindsay R.C. *J. Dairy Sci.* **1969,** *52 (7),* 1097-1100.

(7) Schieberle P.; Grosch W. *J. Agric. Food Chem.* **1987,** *35,* 252-257.

(8) Guth H.; Grosch W. *Lebensm. Wiss. Technol.* **1990,** *23,* 513-522.

(9) Dollmann B.; Wichmann D.; Schmitt A.; Koehler H.; Schreier P. *J. AOAC Intl.* **1996,** *79,* 583-586.

Chapter 6

Tequilas: Charm Analysis of Blanco, Reposado, and Anejo Tequilas

Mercedes G. López[1] and Jean Pierre Dufour[2]

[1]Unidad de Biotecnología e Ingeniería Genética de Plantas, Centro de Investigación y de Estudios Avanzados del IPN, Unidad Irapuato, Apartado Postal 629, Irapuato, Gto. 36500, México
[2]Food Science, Otago University, Dunedin, New Zealand

Tequila is a Mexican alcoholic beverage obtained from *Agave tequilana* Weber var. azul that requires from 8 to 12 years to mature. Traditionally, raw agave pines are cooked, fermented, and double distilled to produce Blanco, Reposado, and Añejo tequilas. Recently, some important tequila flavor components have been published. The main aims of this study were to establish the differences among the three tequila types and to determine the most potent odorants in each type by gas chromatography-olfactometry. Tequila extracts were obtained by liquid-liquid extraction with dichloromethane. Fifth dilution series was used to determine the charm values. Qualitative and quantitative differences were observed in all tequilas. The most potent odorants were phenylethanol and phenylethyl acetate in Blanco tequila; phenylethanol, phenylethyl acetate, and vanillin in Reposado tequila; and phenylethanol, vanillin, and an unknown in Añejo tequila. A good correlation was found between flavor compounds and the aging process of tequilas in white oak casks.

Introduction

Agave plants are cultivated in México for production of different alcoholic beverages such as Mezcal, Pulque, Bacanora, and Tequila, to mention some. However, *Agave tequilana* var. azul is the most cultivated variety of Agave spp. since it is the main raw material used for the production of Tequila. *A. tequilana* is cultivated in México mainly in the states of Jalisco, Guanajuato, Tamaulipas, Michoacán, and Nayarit. A large percentage of the tequila production is exported and 65% of this amount is exported to the Unites States.

Traditionally, *A. tequilana* plants are harvested between 8 and 12 years, and the pines (cores) are cooked in brick ovens for at least 32 hours at 100 °C, the agave juice is then fermented for 24 hours and finally, a double distillation is performed to produce Blanco (white) or Plata (silver) tequila (Figure 1) (1). Tequila Blanco can then be matured in large wood tanks from two to six months to produce Reposado type or in white oak barrels from six to 24 months to generate Añejo class. Tequila production is regulated by the Official Norm NOM006-SCFI-1993. After aging most tequilas are diluted with demineralized water to adjust the proof entry to 110 or 40 proof. Very few studies have been carried out on tequila flavor and aroma. Benn and Peppard (2) found the most potent odorants in some tequilas by instrumental and sensory analysis. In 1999, López (3) published some of the main volatiles involved in Tequila aroma.

Objectives

Therefore, the main objectives of this study were to determine the most potent odorants in Blanco, Reposado, and Añejo tequilas from the same batch and to establish the main differences among these three tequilas.

Materials

Tequila samples were a gift from one of the largest and well known Tequila Companies. Blanco tequila was only distilled, Reposado tequila was distilled and rested for three months and Añejo tequila was distilled and aged for a year.

Experimental

Figure 2 shows the methodology employed to obtain the extracts from all the tequilas samples. One hundred mL of tequila samples (Blanco, Reposado, and Añejo) were diluted with 100 mL of distilled water and then extracted three times with 50 mL CH_2Cl_2. The extracts were combined and dried over anhydrous Na_2SO_4 column. The extracts were concentrated in a Kuderna-Danish apparatus with a nitrogen flow (10 mL/min) to a 100 μL volume. A gas chromatography coupled to olfactometry (GC-O) was used to generate the charm for the characterization of the compounds present in each of the extracts. The chromatographic conditions used were: a Nukol (25 m X 0.25 mm) column with an oven temperature of 45 °C, for 5 min, at a rate 1 of 20 °/min to 100 °C for 1 min and rate 2 of 3 °C/min to 190 °C for 40 min. The injector and detector temperatures were 200 and 220, respectively. The Kovats indeces were calculated using a hydrocarbon series (C_8-C_{26}) from Aldrich.

Fifth dilution serial (1/5, 1/25, 1/125, 1/625, and 1/3125) was used to establish the charm values of the most potent odorants in each tequila.

The characterization of all active compounds was performed in a GC-MS under the same chromatographic conditions as above, except for the length of the column which was 60 m long in this case.

Cooking

Distillation

Jima

Fermentation

Agave plant

Milling

Blanco Tequila

Maturation

Tequilas

Figure 1. Tequila process.

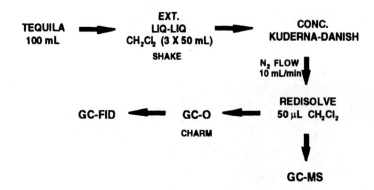

Figure 2. Diagram of the methodology used to prepare the Tequila extracts.

Results and Discussion

The main differences among the analyzed tequilas were essentially quantitative more than qualitative. Figure 3 shows a typical aromagram/chromatogram profiles of a Blanco tequila extract.

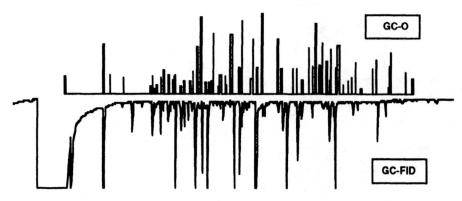

Figure 3. Analysis of Blanco tequila extract by GC-O (aromagram) and GC-FID (chromatogram).

On the other hand, Figure 4 shows the original aromagrams of all tequilas. It is important to mention than the main differences are observed at the beginning of the aromagrams, therefore, more highly volatiles compounds are present in Reposado and Añejo tequila classes that in the non aged tequila. However, many odorants are common to all samples. Table I lists most of the volatiles that were completely characterized in all the tequila extracts.

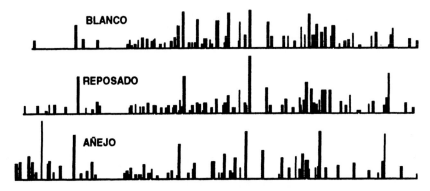

Figure 4. Aromagrams of the original tequila extracts.

On the other hand, the dilutions are shown in Figures 5 (Blanco), 6 (Reposado), and 7 (Añejo). In spite of the large similarity of odorants in all tequilas, it is clearly seen that the most potent odorants in each tequila are quite different not only in number but also in intensity. Three active odors were present in the last dilution series (1/3125) of Blanco tequila. On the other hand, reposado tequila presented six potent odorants responsible of its flavor and Añejo tequila only displayed two very relevant odorants. From this data , it can be said that the Reposado tequila has a more complex overall aroma than the other two classes, in spite of the large difference in aging time between Reposado (two months) and Añejo (six months).

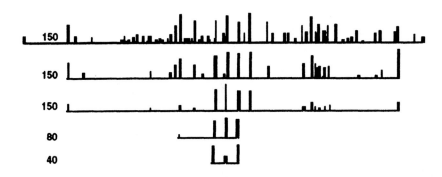

Figure 5. Dilution series of tequila Blanco.

It is worthwhile to mention that the three most potent odorants in Blanco tequilas are also very important in the Reposado class but not so much in the Añejo. However, one of the two most potent odors in Añejo tequila is also potent in the Reposado class.

Table I. Main Volatiles in the Three Classes of Tequila

Group Compounds	KI	Blanco	Reposado	Añejo
ALCOHOLS (21)				
Butanol, 3-methyl	1197	X	X	X
4-Penten-1-ol	1282	X	X	X
2-Heptanol	1293		X	X
Pentanol, 4-methyl	1298	X		
Hexanol	1335	X	X	X
4-Heptanol	1351	X	X	X
Propanol, 3-ethoxy	1363	X	X	X
4-Hexen-1-ol	1373	X	X	
3-Octanol	1379	X	X	
Cyclohexanol	1394	X	X	X
Pentanol, 3,4-Dimethyl	1412	X	X	X
Hexanol, 2-ethyl	1480	X	X	X
2-Nonanol	1509	X	X	
Octanol	1549	X	X	X
Undecanol	1747	X		
Benzyl alcohol	1872			X
Phenylethyl alcohol	1904	X	X	X
ESTERS (40)				
Propanoic acid EE	1175		X	X
Hexanoic acid EE	1217	X	X	X
Ethyl lactate	1330	X	X	X
Octanoic acid EE	1421	X	X	X
Pentanoic acid 4-oxo EE	1601	X	X	X
Butanoic acid 2-hydroxy	1605	X		
Decanoic acid EE	1622	X	X	X
Ethyl succinate	1667	X	X	X
Formic acid 2-phenyl EE	1773	X	X	
Acetic acid 2-phenyl EE	1799	X	X	
Dodecanoic acid EE	1824	X	X	X
Tetradecanoic acid EE	2029			X
Hexadecanoic acid EE	2235	X	X	X
Ethyl 9-hexadecanoate	2259	X	X	X
Octadecanoic acid EE	2354	X	X	
Ethyl oleate	2442	X	X	
Linoleic acid EE	2486	X	X	X
Octanoic acid phenyl EE	2597		X	X
ACIDS (13)				
Propanoic acid 2-methyl	1179	X	X	
Acetic acid	1468	X	X	X
Propanoic acid	1553	X	X	X
Butanoic acid	1639	X	X	X

Table I. (Continued)

Group Compounds	KI	Blanco	Reposado	Añejo
Pentanoic acid	1747	X		
Hexanoic acid	1852	X	X	X
Octanoic acid	2065	X	X	X
Decanoic acid	2276	X	X	X
Dodecanoic acid	2471	X	X	X
Pentanoic acid 4-hydroxy 4-methyl-γ-lactone	1628	X	X	X
Octanoic acid 4-hydroxy γ-lactone	1940		X	X
TERPENOIDS (24)				
Linalool	1540	X	X	X
Terpin-4-ol	1590	X	X	X
Ocimene	1653			X
Linalyl propanoate	1686	X	X	X
Neral	1720	X	X	
t-Linalool oxide	1728	X	X	X
Menth-2-en-7-ol	1743	X		
Citronellol	1757	X	X	X
Nerol	1839	X	X	
2-Hydrocineole	1845	X	X	
t-Geraniol	2004	X	X	X
Nerolidol	2027	X	X	
Cymen-9-ol	2094	X	X	X
Eugenol	2157	X	X	X
Guaiol	2198	X	X	X
Farnesol acetate	2244	X	X	
Farnesol	2340	X	X	
FURANS (12)				
Furan	1456	X	X	X
Furfural	1468	X	X	X
2-Acetylfuran	1504		X	X
5-Methyl-2-furfural	1570	X	X	X
2-Acetyl-2,5-Dime-furan	1573	X	X	X
2-Furanmethanol	1662	X	X	X
2-Acetyl, 5-mefuran	1856	X	X	X
KETONES (7)				
3-Heptanone, 2,4-Dime	1456	X		
3-Hexen-2-one, 5-Me	1484	X		
3-Hexanone-2-methyl	1504	X		
2-Heptanone, 5-methyl	1570	X	X	X
Cyclohexanone	1573			X
2-Hepten-4-one, 2-Me	2010	X	X	X
4-Heptanone, 3-ethyl	2023	X		

Continued on next page.

Table I. (Continued)

Group Compounds	KI	Blanco	Reposado	Añejo
PHENOLS (9)				
Phenol, 4-methyl	2087	X	X	
Tymol	2173	X	X	
Phenol, 4-ethyl	2181	X	X	X
Carvacrol	2189	X	X	
Cresol	2206	X	X	X
Phenol	2263			X
OTHERS (50)				
Propanal, 3-ethoxy	1175	X	X	
Benzaldehyde	1516	X	X	X
Vainillin	2561	X	X	X
Syringaldehyde	2650		X	X
Ethyl vanillate	2661			X
Vainillin derivative	2674		X	X
Thiazole, 2-ethoxy	1667	X		
Thiopene, 2-methyl	1890	X	X	X

NOTE: KI stands for Kovats index; EE for ethyl ester; Me for methyl; Dime for dimethyl.

Figure 8 presents the Charm chromatograms of the three tequilas. Here, the differences among all tequilas are based on the integration of all the dilution series. It is obvious that Reposado and Añejo tequilas have a more complex aroma profile than Blanco tequila. This is certainly due to the resting and aging processes. Again, there are many common aromas in all samples, however, the intensity (Charm values) of many odorants are superior for Reposado and Añejo.

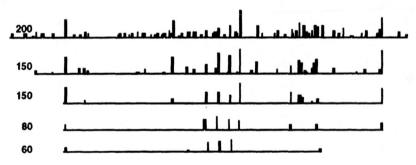

Figure 6. Dilution series of tequila Reposado.

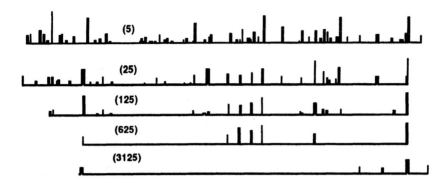

Figure 7. Dilution series of tequila Añejo.

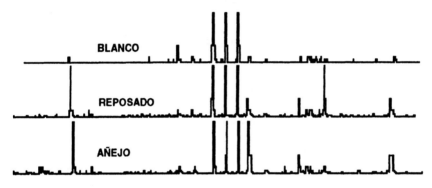

Figure 8. Charm response chromatograms of Blanco, Reposado, and Añejo Tequila series.

Finally, Table II lists the most potent odorants in all tequilas along with the Kovat indeces and the descriptors given to each odorant. Based exclusively on the Charm values of the odorants listed in Table II, it is very clear that the aroma compounds present in the Añejo class have much higher odor values.

Conclusion

It can be concluded that the Añejo class has a more complex overall aroma. Most of these odorants must be generated during the aging process in white oak barrels. However, it is also important to mention at this point, that three of these potent compounds are still unknown, therefore, it is crucial to carry out more experiments using larger tequila volumes to be able to charcaterized these compounds by GC-MS.

Table II. Most Potent Odorants in all Tequilas

KI	Compounds	Descriptor	Charm values		
			Blanco	Reposado	Añejo
1030	Unknown	Solvent	748	845	2842
1200	Butanol, 3-methyl	Alcohol, vinous	2407	2065	6515
1659	Decanoic acid ethyl ester	Fatty	267	400	357
1809	Phenylethyl acetate	Tepache, floral	1564	2415	3035
1862	Unknown	Medicinal	880	1501	2221
1906	Phenlylethyl alcohol	Sweet, floral	6083	4560	7771
1953	Unknown	Plastic		1644	16956
2166	Eugenol	Medicinal, sweet	941	1498	2403
2201	Terpenoid	Chicken	1259	2241	4733
2266	Decanoic acid	Fatty		411	2102
2555	Vainillin	Vainilla, sweet	1959	3641	5510

References

1. Cedeño, M. *Crit. Rev. Biotechnol.* **1995**, *15*, 1-11.
2. Benn, S.M.; Peppard, T.L. *J. Agric. Food Chem.* **1996**, *44*, 557-566.
3. López, M.G. *Flavor Chemistry of Ethnic Foods* ; Shahidi, F.; Ho, J.C., Eds.; Plenum Publishing Co., **1999**; pp 211-217.

Chapter 7

Aroma Active Internal Standards for Gas Chromatography–Olfactometry of Grapefruit Juices

Russell Rouseff[1], Prashanthi Jella[2], Rusty Bazemore[3], and Jing-jing Yang

Citrus Research and Education Center, University of Florida, Lake Alfred, FL 33850
[1]Corresponding author
[2]Current address: Coca-Cola Company, Corporate R&D, Atlanta, GA 30301
[3]Current address: Monell Chemical Senses Center, Philadelphia, PA 19104

The use of internal standards is a classical analytical technique employed to improve quantitative accuracy. In this study, internal standards were added to grapefruit juice solvent extracts to improve the qualitative identification of aroma active compounds. In GC-Olfactometry the human response is ultimately compared to an instrumental response such as that from a mass spectrometer or flame ionization detector (FID) to identify the component responsible for the aroma activity. Using two internal standards which have both olfactometry and instrumental responses, it is possible to exactly align both aromagrams and chromatogram to reduce the uncertainty in assigning the instrumental response associated with aroma activity. The purpose for this study was to develop a set of aroma active components for citrus juices that would have unique sensory qualities, compatible retention times and produce measurable sensory and instrumental responses that could be used to synchronize both instrumental responses and aroma activity. Identification accuracy was improved when both early and late eluting aroma standards were employed. Benzaldehyde, methyl jasmonate and S-methylthio butanoate are compounds that can perform satisfactorily with citrus juice aroma extracts as aroma active internal standards.

Flavor is undoubtedly the single most important factor in the consumer's marketplace and consumption choices and has been extensively studied by both sensory and analytical chemistry scientists. Sensory scientists are primarily interested in how humans perceive or respond to a certain food or aroma sample. They care little about chemical composition. Analytical chemists are primarily interested in a product's chemical composition and devising ways to measure individual components.

They generally assume that the human response to a product will be governed by its total composition. However, it is generally accepted that not all chemical components have flavor activity. This means that if flavor is to be studied intelligently, then techniques that denote which components possess sensory activity must be employed. Gas chromatography - Olfactometry or GC-O, has been the tool of choice to study flavor volatiles because it combines the separation ability of high-resolution capillary gas chromatography with the unique selectivity and sensitivity of the human nose. It has progressed from simply sniffing the hot effluent from the GC and noting the aroma on the recording chart paper to specialized units employing humidified air with computerized data recording devices.

GC-O is used to determine which of the many volatiles that elute from the gas chromatograph have aroma activity (1). For some applications, this is sufficient. However, in many cases it is also necessary to determine peak odor quality and to determine the relative aroma strength of that component. As discussed in recent GC-O reviews (2,3), the overall goal is to identify the set of aroma active components that are responsible for the characteristic flavor of the sample being analyzed. Furthermore, when complex flavor systems are studied, the identification of aroma active components sometimes becomes difficult, because highly potent (low concentration) components will produce significant aroma impact but have little if any instrumental response. The typical GC-O approach to this identification problem is to calculate the unknown component's standard retention index value (Kovat or Ester Index) from the GC-O peak's retention time and use the sensory descriptor to distinguish between the many components which have similar retention index values. Standardized retention index and sensory quality are two independent values which satisfy minimum tentative identification requirements. However, in practice sniffers rarely use the same aroma descriptors as those listed in identification tables. The uncertainty or "fuzziness" in the descriptor makes identification difficult. Analysts must often interpret sniffer aroma descriptors and speculate what they might have meant when they used a particular descriptor. This becomes even more difficult when the preferred practice of using two or more sniffers is employed. Therefore, the common practice is to employ two (or more) dissimilar chromatographic columns which will provide two different sets of retention index values which can be cross compared for identification purposes (4) along with the sensory descriptor(s). However, a careful examination of the literature typically produces a range of standard retention index values for the same compound. This "fuzziness" in retention index values can make identification difficult as it increases the list of candidate aroma impact compounds. The final step in identification is to inject each candidate compound and authenticate its retention index behavior and sensory attribute for each sniffer. Due to the uncertainty in retention index and descriptors, the list of candidate aroma active compounds which must be individually tested can be daunting.

One means of reducing the list of candidate compounds is to employ specialized detectors in place of the traditional FID detector. These include selected ion monitoring mass spectrometry (SIM-MS) and/or sulfur/nitrogen specific detectors. Whereas these detectors provide additional unique information to the analyst, they are also responsible for additional problems. The major problem with these detectors is that they operate under partial vacuum which has the effect of altering retention times. Samples are sucked rather than pushed through the capillary column.

External standards have been used as a standard analytical technique for many years. Their primary use has been to improve quantitative accuracy in samples which have matrix problems, to correct for extraction efficiencies and to compensate for differences in concentrating and injection volumes. However, they can also be used as qualitative markers when comparing retention behavior in different chromatographic or detection systems.

The purpose for this study was to determine if the use of aroma active internal standards could be employed to more precisely synchronize GCO aromagrams with FID, MS, sulfur specific (sulfur chemiluminescence detection, SCD) chromatograms thus reducing the list of candidate aroma active compounds needed to be tested to identify the aroma active components in citrus juices.

Materials and Methods

Samples and chemicals

Commercial not-from-concentrate orange and grapefruit juice were used as samples without modification. Gas chromatographic grade extracting solvents were purchased from Fisher Scientific (Pittsburgh, PA). Flavor standards were purchased from Aldrich (Milwaukee, WI).

Sample Preparation

Samples were prepared using liquid-liquid extraction in the same manner as *(5)* except pentane: ether (1:1) replaced methylene chloride as the extracting solvent. Internal standards (Benzaldehyde, methyl jasmonate and/or S-methylthio butanoate) were added to the extract prior to extract volume reduction.

GC-olfactometry

A HP-5890A GC (Palo Alto, CA) with a standard FID detector was employed to separate juice extracts. The second detector base was modified into an olfactometer by Datu (Geneva, NY) similar to the design and experimental conditions described by Acree and coworkers *(1)*. A Gerstel (Baltimore, MD) column splitter was employed to split the column effluent. Connecting capillary tubing diameters and lengths were selected to split about 75% of the column flow to the sniffer and about 25% to the FID detector. In a few experiments, the column was connected directly to a single detector.

Chromatographic Conditions

Separation was accomplished on a 30 m x 0.32 mm x 0.5 μm film thickness low-bleed DB-5 column (J&W, Folsom, CA). Column oven temperature was programmed from 35 to 275°C at 6°C/min. with no holds. Injection was splitless. All other conditions were the same as *(5)*.

Osme analysis

The Osme, time-intensity approach developed by McDaniel and coworkers *(6)* was used to evaluate odor quality and intensity at the sniff port. Sniff panelists consisted of two males and one female. Training consisted of practice runs with a standard mixture of eleven components typically found in citrus juices (ethyl butanoate, cis-3-hexenol, trans-2-hexenal, alpha pinene, myrcene, linalool, citronellol, carvone, terpin-4-ol, geranial, and neral). Analysis of standards served to familiarize panelists with the Osme aroma intensity scale, optimum positioning and breathing technique, and to provide practice with verbal descriptors.

Sulfur Analysis

Sulfur components from grapefruit juice extracts were volatilized using a special Gerstel (Baltimore, MD) cool injection system, model CIS-3. Samples were injected with an initial injector temperature of 40°C to minimize the possibility of creating thermal artifacts from highly reactive sulfur compounds. Injector temperature was immediately increased at 12°C/sec to a final temperature of 180°C. The same column and temperature program described under chromatographic conditions was employed. Detection was accomplished with a sulfur chemiluminescence detector, SCD, Sievers Instruments (Boulder, CO).

Results and Discussion

Limits of capillary gas chromatography

High resolution gas chromatography has been one of the most effective tools for studying complex aroma volatiles because of its ability to separate hundreds of volatile components in a relatively short time. However, even with this tremendous resolution power, the probability of co-eluted and unresolved components is not generally appreciated. As shown in Figure 1, Martin and coworkers *(7)* were able to demonstrate that the probability of resolving all the peaks in a theoretical mixture increased as peak capacity (column efficiency /theoretical plates) increased. However even with a column with a peak capacity of 2000 there was only about a 65% probability of resolving all peaks in a 30 component mixture and less than 10% probability of resolving all the peaks in a 90 component mixture. This has enormous implications when trying to resolve the components in a flavor mixture which typically exceeds 100 components because unresolved peaks will be the rule rather than the exception. Unresolved chromatographic peaks make it more difficult for the analyst to determine which of the unresolved components is actually responsible for aroma activity because at any given retention time there may be more than a single component present. An additional factor which adds to the difficulty of identifying aroma active compounds is that the aroma active component might be a trace component at the base of a flavor inactive component present in much higher concentration. If a mass spectrometer is used to identify the aroma active component, the fragmentation pattern would erroneously imply that the aroma active component

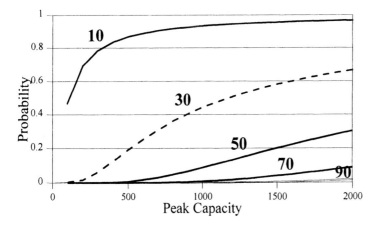

Figure 1 - Probability of separating multicomponent mixtures as a function of column capacity (efficiency). Mixtures range from 10 - 90 components. Data taken from reference (7)

was due to the component in higher concentration, primarily because the mass spectrum was obtained in the region where aroma activity was observed.

Identifications from Standard Retention Index Tables

Standardized retention index values (Kovat/alkane or Ethyl Ester) have been established for the purpose of identifying unknown flavor volatiles *(8-10)*. However, if temperature programming is used, (as is usually the case) then a certain amount of uncertainty will be introduced, as the calculated retention index value will be dependent upon initial column temperature, temperature programming rate, and carrier gas linear velocity *(11)*. Acree and Barnard *(12)* proposed a range of four index units for matching peaks based in part from chromatographic behavior such as peak width at half peak height.

Highly accurate retention index values can be determined only under isothermal conditions. Since most flavor extracts contain compounds of wide volatility range, isothermal analyses at several temperatures would be required. This is rarely done because of time considerations and the relative ease of temperature programmed GC analysis. There are a few published procedures to improve the accuracy of calculated retention index values to within one unit when temperature gradients are employed for nonpolar columns and a "few" units for polar columns taken under optimized chromatographic conditions *(13)*. Unfortunately, the maximum rate of temperature change is only 2 °C/min., which is too slow to complete analyses for most extracts within the maximum time (30 min.) to maintain sniffer attention.

Internal standards of known retention index values can be added to the sample to provide guidance to the analyst in terms of adjusting the calculated retention index to standard index table values. If the alkane standard mixture used to calculate retention index values exactly compensated for different temperature programs, the use of internal standard would be unnecessary. However, polar molecules such as alcohols and esters are highly dependent on column polarity and adsorption activity and very little on column temperature. On the other hand, the retention behavior of hydrocarbons are less sensitive to column adsorption but heavily dependent on column temperature. For example, acenaphthylene and acenaphthene have temperature increments of about 1.0 retention index unit per °C whereas the corresponding values for methyl decanoate and 1-dodecanol are only 0.1 retention unit per °C *(13)*. The use of internal standards to adjust for differences between calculated retention index values and those from a table will be most effective when the internal standard has similar polarity and the same functional group as the component in question. This approach has limited value since in most cases the composition of the peak in question is usually unknown.

Separation of citrus juice aroma extracts

As shown in Figure 2, FID chromatograms of citrus juice extracts are fairly complex. (The first four minutes are not shown as they contain only the void volume and solvent front). Citrus juice or oil chromatograms can be roughly divided into a top note region, a terpene rich section, a carbonyl/ alcohol section and a sesquiterpene rich section. These designations are not exclusive by any means, and are offered only to provide rough organization to the rather complex chromatogram.

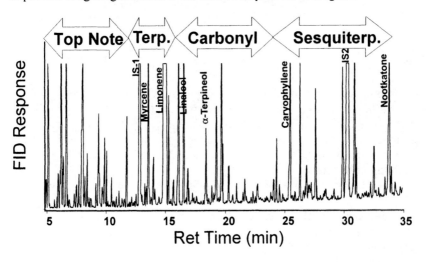

Figure 2 - Chromatogram (DB-5) of grapefruit juice pentane: ether extract containing two aroma active internal standards, IS1 = benzaldehyde and IS2 = methyl jasmonate

The top note section usually includes ethyl butyrate, and hexenal whereas the terpene region includes myrcene and limonene but also octanal. Alcohols and aldehydes predominate in the section labeled carbonyl and include linalool, linalool oxides, and α-terpineol, as well as nonanal and decanal. The sesquiterpene region ranges from caryllophylene to nootkatone and obviously also includes sesquiterpene ketones. It should be pointed out that this chromatogram is from a pentane – ether L/L extraction of grapefruit juice. Chromatograms from dynamic headspace purge and trap aroma extracts are profoundly different, providing a complex topnote fraction but few peaks after limonene are ever observed. Over 130 chromatographic peaks are observed in the chromatogram shown in Figure 2. A detailed examination of these chromatograms indicated that not all peaks were chromatographically resolved and given the discussion in the previous section there is a high probability of co-eluting components.

Determining Aroma Activity

One of the early approaches to determine aroma activity and relative aroma strength was through the calculation of odor units (also called odor activity values *(14)*. Odor units were calculated from the ratio of the concentration of the component of interest divided by its aroma threshold. Since there are typically large differences between odor unit values, they are often displayed as log values. If a component's concentration exceeds its threshold value,

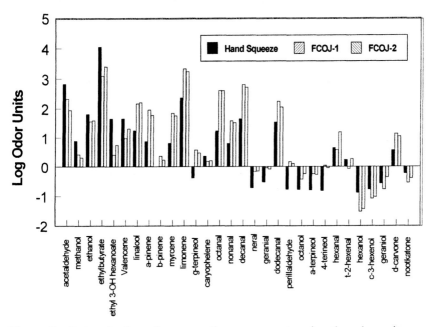

Figure 3 - Determination of aroma active components using the odor unit approach. Data taken from reference (13).

the ratio will be greater than 1.0. Since the log of one is zero, all positive log values indicate that the component should have aroma activity. Shown in Figure 3 is an example of this approach using orange juice data taken from Wade and coworkers *(15)*. According to this approach, the most intense aroma active components are ethyl butyrate, limonene, octanal, decanal, acetaldehyde and dodecanal. The weakness in this approach is that aroma threshold values should be determined in the matrix in which the sample exits. However, aqueous threshold values are usually used, as they are usually more readily available. Since matrix effects can appreciably raise threshold values, odor unit values based on aqueous thresholds are inflated, but depending on the matrix, more for some compounds than others. It is worth noting that recent investigators *(16)* have attempted to overcome matrix effects by measuring head space volatile concentrations to calculate odor units, and have compared their results with those obtained from GC-O aroma dilution (AEDA) results from red wine. However, relative strengths of the major aroma impact compounds from the two approaches were appreciably different.

In their study of orange volatiles using GC-O (Charm analysis) Marin and coworkers *(17)*, found that limonene produced very little aroma impact. Although subsequent GC-O studies reported slightly higher values for limonene *(18,19)*, it was not the major flavor impact compound as predicted from aroma/ odor unit calculations. The difference is probably due to matrix effects from the juice, which contains high sugars, pectin, pulp and other suspended solids.

GC-O Internal Standards

Internal standards have been employed as a standard analytical technique to: correct for matrix problems, determine extraction efficiencies, and in the case of GC, to compensate for injection differences as well as extract volume reduction differences. The concentration of the compound of interest is determined from a response factor calculated from peak areas from standards relative to peak areas from the sample. (Assuming constant injection volumes). The internal standard should be as similar as possible to the component of interest. It should elute close to the peak of interest and have similar instrumental responses.

Perhaps the optimal internal standard is an isotope of the compound of interest. The technique called isotope dilution assay, IDA, was first used to determine glucose in plant tissues *(20)* and has become a classical analytical technique. Schieberle and Grosch pioneered the use of IDA in flavor analysis when they quantified two potent aroma active heterocyclic nitrogen compounds in bread crust. Grosch and coworkers *(21,22,14,23,24)* have since applied stable isotopes in analyzing a variety of food products. One requirement is that the analysis requires a mass spectrometer but the major limitation is the lack of commercial sources of isotopes for many components of interest.

The quantitative internal standard technique requires that the peak of interest be known. We propose to use internal standards to help identify the aroma active peak of interest. GC-O internal standards should have the following characteristics:
1. substantial aroma activity with an equivalent instrumental response.
2. the aroma quality should be dissimilar from the sample components
3. should not pose a health risk to the human assessors
4. should not co-elute with sample components of interest

Since the objective in using these internal standards is to more accurately synchronize aromagrams with other instrumental responses, it would be better to have two internal standards, one which eluted early in the chromatogram and one which eluted towards the end of the chromatogram, bracketing the components of interest.

Retention time shifts

Sometimes it is possible to identify the aroma active component only from its aroma quality and calculated retention index value on one or two chromatographic stationary phases. This is possible when the aroma quality of the unknown component is unique, such as vanillin. Other descriptors such as burnt sugar or floral are more difficult and require additional information that can be obtained from a MS or SCD detector. The difficulty with both of these detectors is that they operate under partial vacuum and will have lower retention times (and calculated retention index values) for the same compound assessed at the sniff port at atmospheric pressure. Some typical data is shown in Table 1 for a group of aroma active sulfur compounds. These compounds were obtained from a whole text search of the Journal of Agricultural and Food Chemistry in 1998 using the search keys of grapefruit and sulfur or fruit and sulfur.

Table 1. – Comparison of retention behavior associated with detector types.

Compound	RT (FID, min.)	RT (SCD, min)	Shift (min)
3-mercaptohexanal	6.03	5.63	-0.4
3-mercaptohexanol	7.21	5.82	-1.39
4-mercapto-4-methylpenetan-2-one	7.53	5.89	-1.64
4-mercapto-4-methylpenetan-2-ol	12.32	8.67	-3.65

Retention time data were obtained from a single GC system (constant pressure) in which the end of the capillary column was simply switched from the FID detector to a SCD detector. Retention times for the same compound were reduced for the SCD detector because this detector operates under a partial vacuum. Shifts in retention time can be calculated if column inlet and outlet pressures are known *(25)*, but the results are approximate at best because several of the equation's components are not accurately known.

It should be pointed out that the retention shift increases as the oven temperature increases (single linear temperature ramp). As the column temperature increases, the viscosity of the carrier gas (helium) decreases, thus retention times are reduced when the column outlet is under vacuum. The data shown in table 1 required 1000x higher concentrations of the sulfur compounds for FID data because of the superior sensitivity and selectivity of the SCD detector. Without a sulfur containing internal standard, it would be difficult to adjust the SCD chromatogram with the FID chromatogram or aromagram. Similar shifts can be expected from mass spectrometer based detectors.

One approach to these retention time shifts is to employ a splitter at the end of the column. The length and internal diameter of the two connecting capillary tubing are selected to produce similar retention times at the sniff port and the instrumental detector. This approach is more successful for instruments based on constant pressure rather than constant flow.

Grapefruit Juice GC-O Studies

If the information from an instrumental detector is to be used to identify an aroma active component, then the synchronization of the instrumental detector, with the olfactometer response is critical. Synchronization may be accomplished by one of two means. If the GC column is not split and the end of the column simply changed to the alternate detector, then the instrumental response at the same time (or retention index value) where the odor of interest was observed is used to identify the unknown component. It must be assumed that changing the end of the column from one detector to another does not change the retention time of the component of interest or if there was a change, it could be detected. As we have seen from the data in Table 1, this assumption is not always warranted.

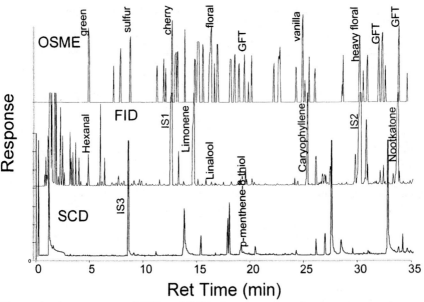

Figure 4 - Aromagram and FID and SCD chromatograms from a pentane-ether grapefruit juice extract. Chromatographic conditions in text. GFT = grapefruit.

Shown in Figure 4 are three stacked chromatograms from a single grapefruit juice extract containing three internal standards. The top trace is the aromagram from Osme. Three internal standards (IS1 = benzealdehyde, IS2 = methyl jasmonate, IS3 = s-methylthiobutanoate) were used in this sample which produced the respective sulfur, cherry and heavy floral olfactory responses. All three internal standards produced substantial instrumental and sensory responses which were used to align the chromatograms. It can be seen that some of the strongest aroma impact compounds (not including the internal standards) are due to hexanal, linalool, vanillin and nootkatone. It is also worth noting that limonene (or an impurity) on its trailing shoulder produced a small aroma peak and caryophyllene had no detectable aroma activity. Interestingly, the tiny peak preceding caryophyllene was due to vanillin, which produced an intense vanilla aroma. As has been previously speculated (4), vanillin probably does not contribute that much to food flavor, because its normal vapor pressure is so low.

There were three peaks with a distinct grapefruit aroma (labeled GFT). The first was from the thiol, p-menthene-8-thiol first discovered by Demole and coworkers (26), the last is from the sesquiterpene ketone, nootkatone. The peak preceding nootkatone by about 3 min. may be 8-9 didehydronootkatone also discovered by Demole and coworkers (27).

Perhaps hardest to align was the sulfur chromatogram. S-methyl thiobutanoate was the third internal standard which produced a sulfur note on the aromagram. We used a single standard because at the time we thought the retention time shift would be consistent, but as shown on Table 1 this is not true. This points out the limitation of the Kovat's or Ester retention index system. These standardized retention index systems work well when the component of interest is measured with the same detector (FID or MS) used to measure the respective alkane or ester standards. However, the sulfur detector does not respond to either alkanes or esters. If one simply determined the retention times for a series of alkanes and simply switched the end of the column. We are currently seeking a late eluting sulfur standard that does not interfere with any of the existing aroma peaks in grapefruit that would provide the necessary second calibration peak.

Shown in Table 2 are the identifications from the averaged GC-O panel results. Identifications were based on matching Kovat's retention index values with those in the literature and also considering the similarity of sensory responses. These identifications should be considered tentative because only a single chromatographic system was employed. A more certain identification can be achieved when retention index values form at least two different chromatographic systems are used. A recent publication on "Fresh" grapefruit juice (28) reported 37 compounds using retention indices from three chromatographic systems to identify components. Even though samples in the current study were heated (pasteurized) and extracted in a different way, there was substantial agreement (17 compounds) between the two studies.

The olfactory retention index values for the early eluting benzaldehyde and late eluting methyl jasmonate serve to improve identification of unknown peaks by serving as reference points when making identifications to suggest that the analyst look to slightly higher or lower retention index values. Because of the wide range in reported retention index values this works best when comparing values from a single source. Even though there is some disagreement in K.I. values for benzaldehyde, it is a suitable standard because it elutes relatively early and does not seriously interfere with grapefruit juice components. Methyl jasmonate also offers a unique aroma, but exists in several forms which distracts from its value as an internal standard.

Table 2. – Average Identification results compared to literature values.

Obser. K.I. DB-5	Observed Descriptor	Identification	Literature K.I. DB-5	Literature Descriptor
722	fruity	methyl butanoate	724 *(9)*	floral/pineapple *(29)*
728	fruity			
801	green	Hexanal	798*(30)*, 804*(31)*	Green, *(30)*, Grass *(31)*
842	fruity	ethyl-2methyl butanoate	849*(31)*	fruity*(31)*
850	roasted grain			
862	oatmeal, mushroom	1-heptene-3-one	873*(10)*	Metallic *(10)*, geranium *(28)*
885	cooked oat			
934	cat urine ,fruity	4-mercapto-4-methyl-2-pentanone	944 *(31)*, *(32)*	Boxtree *(31)*, black currant *(32)*
965	Cherry, almond	Benzaldehyde (IS1)	961 *(9)*, 968*(10)*	Cherry, almond *(10)*
981	Unripe mango	1-octene-3-one	979 *(10)*	Mushroom *(10)*
984	Sweet fruity			
989	minty	Myrcene	991 *(9)*, 992 *(10)*	Balsamic *(10)*, metallic, herbaceous *(33)*
1003	citrus, lemon grass	octanal	999 *(30)* 1001, *(9)* 1006 *(10)*	Orange *(30)*, Soapy *(10)*, chemical, green *(29)*
1026	minty	p-cymene	1026 *(9)* 1027	Solvent *(10)*
1030	citrusy	limonene	1033 *(10)*	citrus *(10)*
1040	citrusy			
1047	floral			
1050	unripe cucumber	(E)-ocimene	1051 *(34)* 1060	Warm, *(10)* green *(35)*
1060	cotton candy	Furaneol	1050 *(36)*, 1065	caramel *(10)*
1072	burnt sugar			
1089	cotton candy, citrusy			
1095	cotton candy	Sotolon	1081 *(31)*, 1107 *(10)*	Curry *(31)* cotton candy *(10)*
1098	burnt sugar, fruity	Nonanal	1102 *(30)*	Fruity *(30)*
1101	floral	Linalool	1097 *(37)* 1103 *(31)*	Flowery, *(38)* Sweet *(37)*, muscat *(31)*
1112	cooked oat			
1120	cooked rice,			
1130	terpeney	Ethyl-3-hydroxyhexanoate	1134* *(28)*	Fruity *(28)**
1148	greenish, floral	Z-2-nonenal	1142 *(30)*	Fat, green *(30)*
1165	rubber, vinyl			
1176	terpeny	terpinen-4-ol	1177 *(39)*,	Terpey *(33)*,Musty *(10)*,

			1182 *(10)*	woody *(38)*
1201	cilantro	a-terpineol	1195 *(10)*	Pine oil *(10)*, Flowery *(33)*
1225	dead bug, terpeney	Decanal	1207 *(10)*, 1210 *(31)*	Soapy, *(10)*, green wood *(31)*, citrus, fatty *(33)*
1269	floral	2-phenylethyl acetate	1260 *(31)*	Rose *(31)*
1299	sulfury, grapefruit	p-menthene-8-thiol		
1312	rancid oil	(E,E)-2,4-decadienal	1311*(30)*, 1317 *(10)*	fatty, tallowy *(30)*, fried *(10)*
1318	spicy, oily	4-vinylguaiacol	1324 *(40)*	Clove *(40)*
1350	coal, smoky			
1384	cooked, caramelized			
1404	vanilla	Vanillin	1410 *(10)*, 1412 *(40)*	Vanilla *(10) (40)*
1492	floral			
1638	Heavy floral	Methyl jasmonate	1614 *(10)*, 1647 *(9)*	Jasmine *(10)*
1708	peppery			
1722	rotten gft.			
1790	grapefruit	Nootkatone	1814 *(28)**, 1847 *(34)*	Grapefruit *(41)*

* = SE-54 retention index value

Conclusion

Three internal standards allow chromatographic responses from different detectors to be compared because they offer a common point of reference in comparing disparate responses. These standards were not chosen at random, but carefully selected by searching GO-O data bases for compounds which would meet the four criteria listed earlier. As can be seen in Figure 4, the choices are reasonable for grapefruit juices. Orange juices produce similar results.

Acknowledgements

The authors gratefully acknowledge the financial support of the Florida Department of Citrus. Florida Agricultural Experiment Station Journal Series No. R-07688

Literature Cited

1. Acree, T. E.; J., B.; Cunningham, D. G. *Food Chem.* **1984**, *14*, 273-286.
2. Grosch, W. *Trends Food Sci. Technol.* **1993**, *4*, 68-73.
3. Mistry, B.; Reineccius, T.; Olson, L. K. In *Techniques for Analyzing Food Aroma*; Marsilli, R., Ed.; Dekker: New York, 1997; pp 265-292.
4. Acree, T. E. *Anal. Chem.* **1997**, *69*, A170-A175.
5. Jella, P.; Rouseff, R.; Goodner, K.; Widmer, W. *J. Agric. Food Chem.* **1998**, *246*, 242-247.
6. Da Silva, M. M. A. P.; Lundahl, D. S.; McDaniel, M. R. In *Trends in Flavor Research*; Maarse, H., van der Heij, D. G., Eds.; Elsevier Science Publishers: Amsterdam, 1994; Vol. 35, p 518.
7. Martin, M.; Herman, D. P.; Guiochon, G. *Anal. Chem.* **1986**, *58*.
8. Jennings, W. G.; Shibamoto, T. *Qualitative analysis of flavor and fragrance volatiles by glass capillary gas chromatography*; Academic Press: New York, 1980.
9. Adams, R. P. *Identification of essential oil components by gas chromatography/ mass spectroscopy*; Allured Publishing: Carol Stream, 1995.
10. Acree, T. E.; Arn, H.,Flavornet,**1997** *http://www.nysaes.cornell.edu/flavornet/*
11. Yin, H. F.; Sun, Y. I. *Chromatographia* **1990**, *29*, 39-48.
12. Acree, T. E.; Barnard, J. In *Trends in Flavour Research*; Maarse, H., van der Heij, D. G., Eds.; Elsevier Science Publishers: Amsterdam, 1994.
13. Farkas, P.; Le Quere, J. L.; Maarse, H.; Kovac, M. In *Trends in Flavour Research*; Maarse, H., van der Heij, D. G., Eds.; Elsevier Science: Amsterdam, 1994; p 518.
14. Milo, C.; Grosch, W. *J. Agric. Food Chem.* **1996**, *44*, 2366-2371.
15. Wade, R. L.; Lee, K. Y.; Swaine, R. L.; Myers, T. S.; Burgard, D. R. In *Citrus Processing Present and Future*; Matthews, R. F., Ed.; University of Florida, Cooperative Extension Service: Gainesville, FL, 1992; pp 181-185.
16. Ferreira, V.; Ardanuy, M.; Lopez, R.; Cacho, J. F. *J. Agric. Food Chem.* **1998**, *46*, 4341-4346.
17. Marin, A. B.; Acree, T. E.; Hotchkiss, J. H.; Nagy, S. *J. Agric. Food Chem.* **1992**, *40*, 650-654.
18. Gaffney, B. M.; Havekotte, M.; Jacobs, B.; Costa, L. *Perfumer & Flavorist* **1996**, *21*, 8-12.
19. Hinterholzer, A.; Schieberle, P. *Flavour Frag. J.* **1998**, *13*, 49-55.
20. Sweeley, C. C.; Elliot, W. H.; I., F.; Rhyage, R. *Anal Chem* **1966**, *38*.
21. Sen, A.; Laskawy, G.; Schieberle, P.; Grosch, W. *J. Agric. Food Chem.* **1991**, *39*, 757-759.

22. Guth, H.; Grosch, W. *Journal of Agricultural and Food Chemistry Germany.* **1994**, *42*, 2862-2866.
23. Semmelroch, P.; Grosch, W. *J. Agric. Food Chem.* **1996**, *44*, 537-543.
24. Pfnuer; T. Matsui; Grosch; H. Guth; Hofmann, T.; Schieberle, P. *J. Agric. Food Chem.* **1999**, *47*, 2044-2047.
25. Cramers, C. A.; Leclercq, P. A. *CRC Critical Reviews in Analytical Chemistry* **1988**, *20*, 117-147.
26. Demole, E.; Enggist, P.; Ohloff, G. *Helv. Chim. Acta* **1982**, *65*, 1785-1794.
27. Demole, E.; Enggist, P. *Helv. Chim. Acta* **1983**, *66*, 1381-1391.
28. Buettner, A.; Schieberle, P. *J. Agric. Food Chem.* **1999**, *47*, 518-5193.
29. Gomes da Silva, M. D. R.; Chaves das Neves, H. J. *J. Agric. Food Chem.* **1999**, *47*, 4568 -4573.
30. Schlüter, S.; Steinhart, H. *J. Agric. Food Chem.* **1999**, *47*, 5146-5150.
31. Kotseridis, Y.; Baumes, R. *J. Agric. Food Chem.* **2000**, *48*, 400-406.
32. Guth, H. *J. Agric. Food Chem.* **1997**, *45*, 3022-3026.
33. Tairu, A. O.; Hofmann, T.; Schieberle, P. *J. Agric. Food Chem.* **1999**, *47*, 3285-3287.
34. Sun, D.; Petracek, P. D. *J. Agric. Food Chem.* **1999**, *47*, 2067 -2069.
35. Ohloff, G. *Riechstoffe und Geruchssinn*; Springer-Verlag: Berlin, Germany, 1990.
36. Sakho, M.; Chassagne, D.; Crouzet, J. *J. Agric. Food Chem.* **1997**, *45*, 883 - 888,.
37. Czerny, M.; Grosch, W. *J. Agric. Food Chem.* **2000**, *48*, 868 -872.
38. Jorgensen, U.; Hansen, M.; Christensen, L. P.; Jensen, K.; Kaack, K. *J. Agric. Food Chem.* **2000**, *48*.
39. Tellez, M. R.; Estell, R. E.; Fredrickson, E. L.; Havstad, K. M. *J. Agric. Food Chem.* **1997**, *45*, 3276 -3278,.
40. Czerny, M.; Wagner, R.; Grosch, W. *J. Agric. Food Chem.* **1996**, *44*, 3268-3272.
41. Berry, R. E.; Charles J. Wagner, J.; Moshonas, M. G. *J. Food Sci.* **1967**, *32*, 75-78.

Chapter 8

SPME and GC–AED–Olfactometry for the Detection of Trace Odorants in Essential Oils

S. Grant Wyllie[1], Kerry-Ann Merry[1], and David N. Leach[2]

[1]Centre for Biostructural and Biomolecular Research, University of Western Sydney, Hawkesbury, Richmond, New South Wales 2753, Australia
[2]Australian Tea Tree Oil Research Institute, Southern Cross University, Lismore, New South Wales 2480, Australia

The presence of odor active trace components in essential oils obtained from *Melaleuca alternifolia* have been investigated using gas chromatography atomic emission detection (GC-AED) and gas chromatography olfactometry techniques (GC-O) in combination with solid phase microextraction (SPME). Distinct qualitative and quantitative differences were found in front of the terpenoid region of both the aromagrams and the chromatograms obtained from four samples of high quality tea tree oil from different sources. These results suggest that the subtle differences observed in the aromas of the sample oils could be caused by the variability of the occurrence of these trace volatiles. The combination of SPME, GC-AED and GC-O is a useful method for studying such problems.

Essential oils have been valued for many centuries for their fragrance, flavoring and medicinal properties. They are usually prepared by extraction of plant material using steam distillation and are not subject to extensive additional processing before further use. The character and quality of the oils is often finally assessed by sensory means in conjunction with specialised gas chromatographic analysis. The Australian native species *Melaleuca alternifolia* was utilized for medicinal purposes by indigenous people and the essential oil became a popular household remedy in Australian homes prior to World War II for the treatment of cuts, abrasions, insect bites and a range of skin conditions. More recently it has become a crop of commercial significance and is increasingly being incorporated into a range of therapeutic and cosmetic products. Tea tree oil as with most essential oils has a distinctive odor which is incorporated into the overall odor of the product. Suppliers of the raw material are therefore required to ensure that the odor of the oils they supply is as consistent as possible. This has caused some difficulties in the past with particular batches of oils taking on notes that are not tea tree oil like. Analysis of these oils using conventional GC does not reveal any differences which would shed light on this problem. This suggests the presence of low odor threshold, trace components which could be envisaged to arise from a number of possible sources such as differences in or contamination of raw materials, differences in extraction protocols and equipment and post extraction handling and storage.

This project investigates the presence of such trace components in a number of oils using an approach based on GC-Olfactometry. In addition because it was anticipated some of these components may be highly volatile the use of SPME as a sample introduction method to overcome any interference from solvents was explored. An atomic emission detector (AED) GC detector was also utilized to provide additional information about the nature of the compounds found. AED is typically up to five times more sensitive to carbon than an FID, more sensitive than GCMS in the scan mode, and ten times more sensitive to sulfur than an FPD (*1*). Its element selectivity is high, a distinct advantage in dealing with the complex mixtures often encountered in food related matrices. With this combination it is possible to simultaneously determine the presence of compounds that contain the elements carbon, oxygen, sulfur, nitrogen together with their odors, in an analytical sample.

Materials and Methods

The GC-AED-Olfactometry System

A Hewlett-Packard G2350A GC-AED was fitted with a GC olfactometry unit (SGE, Vic., Australia). Gas chromatographic conditions were as follows. Injector temperature 260°C, split ratio 1:5. Column: BP5 50m x 0.22μm id, film thickness 1.0μm (SGE, Vic., Australia). Carrier gas helium. Temperature program 60°C for 5 min then 4°C/min to 200°C then 25°C/min to 300°C and hold for 2 min. The standard detector configuration monitored the C193nm and S181nm emission lines. The end of the column was terminated in a two-way splitter from which one fused silica capillary led to the AED and the other to the sniffing port outlet. Because the AED detector operates at a low positive pressure the resistance of the line from the splitter to the olfactory port had to be carefully adjusted so that some sample passes into the detector. This made it difficult to determine the split ratio accurately and a suitable value was arrived at largely by trial and error. In all other aspects the instrument and the GC-O attachment were operated as per the manufacturers recommendations. The odor assessor at the sniff port held a variable DC device whose output was fed directly into one acquisition channel of the chromatography data processing software. By turning a knob the assessor was able to generate time:intensity data in the form of an aromagram which could be overlaid directly on the chromatograms. For this work all odor assessment was carried out by a single individual and each assessment was duplicated.

Static Headspace Analysis System

One drop of oil was placed in a headspace vial (20mL) which was sealed. Headspace sampling was carried out using a Hewlett-Packard 7594 Headspace sampler. The sample was equilibrated at 50°C for 55min. before analysis on a HP6890 GC fitted with a 50m × 0.22mm i.d. fused silica BP-1 column (SGE, Australia). The oven was programmed from 60°C to 270°C at 6°C/min. then held for 2min. Detection was by FID.

Tea Tree Oil Sampling and Analysis Protocol

Samples of four recently processed oils from established producers were obtained and stored at room temperature in glass bottles. For analysis one drop of oil was placed in a brown glass container (7mL) and sealed with a Teflon coated closure. The sample was equilibrated for 45 min before sampling. The headspace was then sampled for 10 min using the SPME fibre and then injected immediately into the GC-AED.

SPME Fibers

SPME fibers used were: Polydimethylsiloxane (PDMS) 100μm; Polydimethylsiloxane/Divinylbenzene (PDMS/DVB) 65μm and Polydimethylsiloxane/Carboxen (PDMS/Carboxen) 75μm. (Sigma-Aldrich, Castle Hill, Australia)

Results and Discussion

The Sample Oils

The four oils were analysed by conventional GC-FID and the results confirmed that these oils conform to the requirements for good quality as defined by ISO 4730 (*2*). While there are small differences (see Table 1) between the concentrations of the fourteen key terpene components monitored by the standard all of these aroma active constituents will be well above their odor thresholds and it is unlikely that such differences would be responsible for the subtle odor differences observed between these oils. Thus while the oils conform to the ISO standard which includes a "characteristic' odor descriptor this subjective measurement is insufficient to enable an accurate appraisal of the odor quality. Such short-comings may be addressed by the use of GC-O to identify important odor impact compounds which can be quantified and incorporated into the standard.

Solid Phase Micro Extraction (SPME) as a sample introduction system

A number of investigators have utilized the strengths of SPME for the analysis of flavors and fragrances and have carried out comparisons between conventional extraction and SPME analyses (*3, 4, 5)*

A range of fiber coatings are available for SPME each having different polarity and sensitivity characteristics.

Coatings evaluated were:

- PDMS (100µm); a good allround fiber for oil headspace but its sensitivity for highly volatile components was only moderate;
- PDMS/Carboxen (75 µm); showed the greatest sensitivity for low boilers but caused rearrangements of some terpenoids and exhibited considerable sample carryover from run to run;
- PDMS/DVB (65 µm); had intermediate sensitivity for low boilers and gave little evidence of sample degradation.

All of these fibers delivered sufficient sample to the analysis system to enable the successful olfactory detection of oil components having comparatively high sensory thresholds such as α-pinene and terpinen-4-ol.

Table I. Composition (relative %) of Major Terpenoid Components of Tea Tree Oil Samples

Compound	Oil D	Oil A	Oil C	Oil B
α-pinene	2.58	2.43	2.22	2.42
Sabinene	0.71	0.71	0.34	0.70
α-terpinene	8.75	8.42	9.66	10.18
p-cymene	2.97	4.42	1.75	1.65
Limonene	0.93	1.05	1.05	0.86
1,8-cineole	2.41	3.45	3.88	2.16
γ-terpinene	20.7	18.9	20.0	20.44
α-terpinolene	3.28	3.27	3.67	3.55
Terpinen-4-ol	41.1	38.9	41.0	37.5
α-terpineol	3.08	3.11	3.16	2.71
Aromadendrene	1.08	1.32	1.06	1.50
δ-cadinene	1.06	1.17	1.02	1.50
Globulol	0.51	0.37	0.44	0.62
Viridiflorol	0.42	0.29	1.37	0.47

Since the different fibers also have specific sample selectivity characteristics a comparison between the oil composition as determined by static headspace and the three fibers was made. Figure 1 shows the composition of the headspace of a sample oil determined by static and SPME headspace using a PDMS/DVB fiber. Somewhat surprisingly given the different sampling procedures the compositions are quite similar indicating, at least for the terpenes, that there is a close correlation between what is sampled by SPME (with this fiber) and what might be encountered in sensory assessment. Figure 2 shows the comparison of the composition of the oil as determined using the PDMS/Carboxen and PDMS/DVB fibers. Again the overall analytical results are similar except in the case of p-cymene and γ-terpinene where the

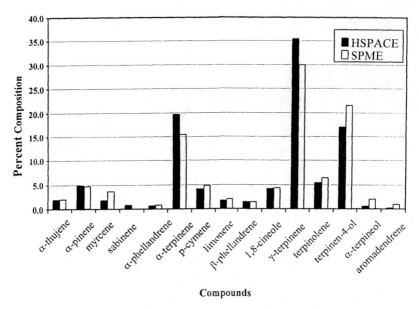

Figure 1. Comparison of the composition of the major components of tea tree oil determined by static headspace and by SPME (PDMS/DVB) of oil headspace.

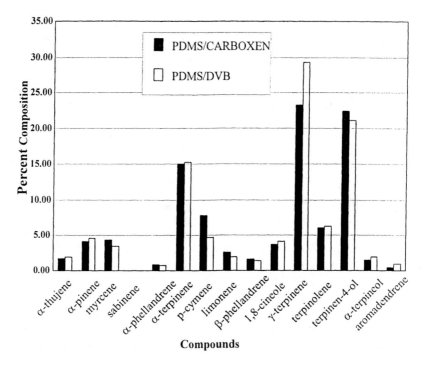

Figure 2. Comparison of the composition of the major components of tea tree oil determined by SPME of oil headspace with PDMS/Carboxen and PDMS/DVB fibers.

rearrangement of the latter to the former on the PDMS/Carboxen fiber is clearly evident. Despite this problem this tip was used for analyses of the more volatile components that elute in the front part of the chromatogram (<20 minutes) where its extra sensitivity for this type of compound could be exploited. To utilize the advantages of each fiber PDMS/Carboxen fibers were used for sampling when the first 20 minutes of the GC-O run were to be evaluated and PMDS/DVB fibers used when samples were injected for odor assessment for terpenoid contributions. It should however be kept in mind that SPME fibers can exhibit specific compound selectivity characteristics which means that components present in the sample may not be introduced into the analytical system. Indeed we have evidence (data not shown) that traces of aliphatic acids present in some tea tree oils are not detected when SPME sampling is used.

GC-O Analysis

An odor assessment of the four oils by experienced analysts showed that each oil was considered to be of good odor quality with no defects but that each had subtly different but characteristic odor notes. Each oil was then analysed using the GC-AED/GC-O system. A typical output is shown in Fig 3. The top trace shows the aromagram incorporating the response from the sniff port assessor, the middle trace shows the S181 response and the lower trace the C193 output, equivalent to that of an FID. Peaks emerging in the first 20 min are non-terpenoid in nature and there are clearly a considerable number of odor active components in this region of the chromatogram which are not being detected by either carbon or sulfur detectors. The only significant peak in the S chromatogram of some of the oils proved to be dimethyl disulfide which despite its low odor threshold (0.16-12ppb) was not always detected by the olfactory analysis. This illustrates the high sensitivity of the S detection system of the AED. The O and N selective detection capabilities of the AED are not as sensitive and for the samples and conditions employed in this study did not provide useful additional information.

A comparison of the GC-O output of the front end of the chromatograms from two of the oils is shown in Figure 4. Arrows indicate peaks in the aromagrams which have little or no response in the corresponding AED trace. It is also apparent that there are small but possibly significant differences between the two oils' GC-O patterns. Similar differences were found between the other oils (data not shown). These results indicate that there are low concentrations of very odor active compounds in this region of the chromatograms and that the qualitative differences between these may contribute to the different notes perceived for the oils.

The identification of these compounds was attempted by GC-MS. The first 20 minutes of the GC-MS trace obtained by the splitless injection of an oil headspace sample using a PDMS/Carboxen fiber is shown in Figure 5. Individually these components generally represent less than 0.2% of the oil but the aromagrams suggest that some of these are likely to contribute to the oil odor. Compounds tentatively identified by comparison of mass spectra and retention indices are ethanol, acetone, ethyl acetate, 2-methylpropanal, 2-methylbutanal, pentan-2-one, ethylfuran, dimethyldisulfide, hexanal and a mixture of hexenols. Of these 2-methylbutanal,

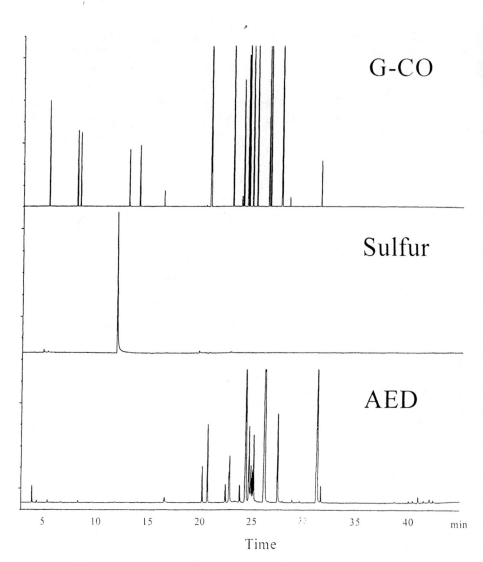

Figure 3. Typical output for a tea tree oil SPME headspace sample from the GC-AED/GC-O system.

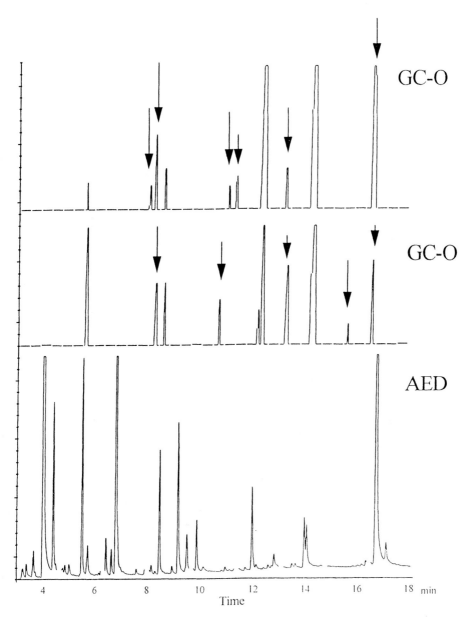

Figure 4. GC-AED chromatogram and GC-O aromagrams for two different tea tree oils from the first 20 minutes of the analysis (PDMS/Carboxen fiber). Arrows indicate GC-O responses which do not coincide with well defined peaks in the AED chromatogram.

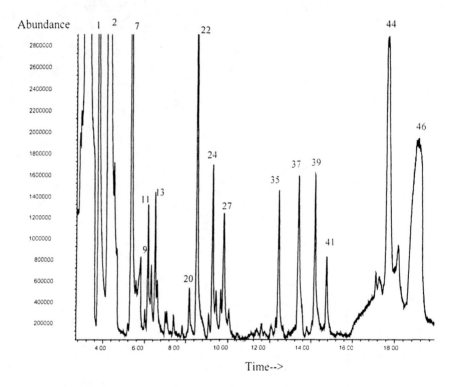

Figure 5. GC-MS trace of the nonterpenoid region of the chromatogram of a typical tea tree oil. Peak numbers are (1) ethanol; (2) acetone; (7) 2-methylpropanal; (11) butan-2-one; (13) ethyl acetate; (22) 2-methylbutanal; (24) pentan-2-one; (27) ethyl furan; (35) dimethyldisulfide; (37) toluene; (39) octane; (41) hexanal; (44) cis-3-hexenol.

dimethyldisulfide (in some cases) and hexanal can be directly correlated with GC-O responses. Some of these compounds are the products of the lipid oxidation pathway *(6)* and presumably are generated during the tissue damage arising from the cutting and distillation process. They bear some resemblance to the volatiles found in olive oils *(7)*. Other GC-O responses were either not detectable or have not yet been reliably identified by GC-MS.

Comparison of this region of the chromatograms of the four oils examined shows (data not shown) that there are small but significant qualitative and quantitative differences between the oils. These differences therefore may well account for the different aroma notes detected in the oils.

A similar aromagram/chromatogram combination is shown in Figure 6 except that the region of the chromatograms from 20-45 minutes has been displayed. The majority of responses in this region are from the terpenes, some of which have low odor thresholds, for example both α-pinene (6ppb) and β-pinene (140ppb) are readily detected by GC-O. There are however some qualitative differences between the GC-O traces and indications of the presence of trace odor actives. Again all four oils showed somewhat different odor responses in this region of the chromatograms. The nature of the compounds responsible for these differences has not been investigated.

Conclusions

This work has shown that there are a number of trace nonterpenoid components having very low odor thresholds present in tea tree oil. These components vary, both qualitatively and quantitatively, between the four oils examined and this may serve to explain the differences observed in their overall odors. The differences may be attributed to variations in the nature of the plant raw material or to differences in the conditions of extraction and subsequent storage. GC-O is a technique that can be used to study the factors which determine these small differences because it has a sensitivity and specificity which current detectors are hard pressed to match. Atomic emission detection can be combined successfully with GC olfactometry and its sensitivity and selectivity should prove useful in this field.

Acknowledgements

The expert technical assistance of W. Eickhoff and L. Diomo is gratefully acknowledged.

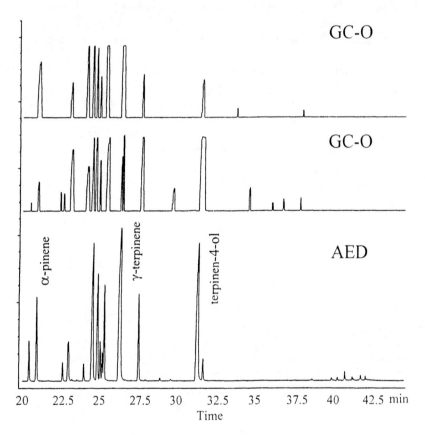

Figure 6. GC-AED chromatogram and GC-O aromagrams for two different tea tree oils over the 20-45 minute region of the analysis (PDMS/DVB).

References

1. Mistry, B.S., Reineccius, G. A. and Jasper, B.L. In *Sulfur Compounds in Foods*; Eds., Mussinan, C.J. and Keelan, M.E., American Chemical Society, Washington DC, 1994; pp 8-21.
2. International Standards Organisation, Oil of *Melaleuca*, Terpinen-4-ol Type, 4730, ISO, Geneva, 1996.
3. Yang, X. and Peppard,T., *J.Agric. Food Chem.* 1994, *42*, 1925-1930.
4. Steffen, A. and Pawliszyn, J. *J.Agric. Food Chem.* **1996**, *44*, 2187-2193.
5. Field, J.A., Nickerson, G., James, D.D. and Heider, C. *J.Agric. Food Chem.* **1996**, *44*, 1768-1772.
6. Hatanaka, A. *Phytochem.***1993**, *34*, 1201-1218.
7. Angerosa, F., DiGiacinto, L. and d'Alessandro, N. *J.High Resol. Chromatogr.* **1997**, *20*, 507-510.

Chapter 9

The Identification of Aroma Compounds in Key Lime Oil Using Solid-Phase Microextraction and Gas Chromatography–Olfactometry

M. G. Chisholm, M. A. Wilson, G. M. Gaskey, J. A. Jell, and D. M. Cass, Jr.

School of Science, Behrend College, The Pennsylvania State University, Erie, PA 16563

The aroma composition of the peel oils from Key lime (*Citrus aurantifolia* Swingle) has been examined for both cold-pressed and distilled samples. The data indicated that loss of some of the more volatile components may have occurred during extraction and distillation, and that many of the odor-active compounds found had not been previously reported.   Solid phase microextraction (SPME), a technique ideal for trapping low concentration volatiles that are difficult to identify, coupled with gas chromatography-olfactometry (GCO) was used to examine the changes in aroma composition of the lime extracts with the extraction method. Three different fibers were used for the extraction, with varying adsorption times, and significant changes in the aroma composition were found using GCO for both the cold-pressed and distilled oils. To identify additional volatiles, the lime oils were concentrated by fractionation and volatiles were identified using SPME-GCO and SPME-GC-mass spectrometry (MS).

Solid phase microextraction (SPME) has become an important and widely used tool for sample preparation in the analysis of volatile mixtures, and is an effective alternative to traditional methods such as liquid-liquid extraction and purge and trap methods for headspace sampling. Its appeal lies in its simplicity, speed and reproducibility, which have resulted in a wide variety of applications. It was first used in water analysis applications (*1,2*) where the SPME fiber was immersed directly into an aqueous sample. It has also been used in a variety of headspace sampling methods where it has been shown to be an effective method for concentrating trace analytes (*3*). In flavor analysis, it has found value as both a screening technique (*4*) and as a tool for concentration of the headspace volatiles (*5*).

Its sensitivity, which can reach the parts per trillion level (6,7) is appropriate for identifying trace odorants with low thresholds. Further, the selection of the fiber used in an analysis can be used to enhance the selectivity for target odorants (8) and several workers have compared the selectivity of different fibers in flavor applications (9). SPME has been found to have many advantages over solid phase extraction (SPE) and purge and trap headspace methods (10,11).

SPME has been applied to the analysis of citrus products by Steffan and Pawliszyn (12) and Jia, et al. (13) who examined the headspace of orange juice and Coleman et al. who used SPME-chiral GC to examine grapefruit oil (14). More recently, SPME has been used in conjunction with GCO for screening odorants in wines (15) and as a quantitative tool for measuring odor intensities (16). This paper describes the use of SPME in an attempt to identify potent odorants in Key lime oil which have eluded detection by GCO and GC-MS using other sample preparation techniques. Extensive lists of the compounds present in Key lime oil have been published (17-20), but no attempt has been made to measure their odor intensities and how much each one contributes to the overall aroma of lime oil.

Preliminary results from the identification of key odorants in cold-pressed and distilled Key lime oil (21) showed that a third of the thirty most potent odorants determined by measuring their odor spectrum values with GCO (22) were not mentioned in the literature. This is shown in Tables I and II. Initially all that was known about these odor-active compounds was that they were present in Key lime oils at above threshold amounts, but below the sensitivity limits of the mass selective detector. Also their retention index (RI) and odor descriptors were determined.

Materials and Methods

Preparation of lime oil. Key limes from southern Florida were purchased at a supermarket and the peel was removed with a zesting tool. For each 10 g of peels, 100 mL of a mixture of pentane:ether (1:1) was added and stirred at room temperature for 1 hour. The mixture was filtered to remove the extracted peels, and the diluted solution of the oil was stored at 5 °C until used. The solvent was removed on a roto-vapor to yield the cold-pressed oil. The distilled oil was prepared by suspending 5 mL of cold-pressed oil in 200 mL of a 5% solution of citric acid in water, heating it under reflux for 0.5 h, then hydrodistilling the suspension until all the oil was collected. Table III shows how the composition of the distilled oil varied with the length of time of the distillation. The distillation conditions above were used for this determination. Samples were prepared for GCO from the pure oil by making a set of serial dilutions with a 3-fold dilution for each sample using a 1:1 pentane-ether mixture. The most concentrated sample was 729-fold and the most dilute was 1/27-fold.

Gas Chromatography-Olfactometry (GCO) of lime oil samples. The samples were analyzed by a single sniffer who had been screened for acuity by using a training procedure described by Marin et al (23). A Charm station was used (Datu,

Table I. Most potent odorants found in cold-pressed Key lime oil.

Peak No.[a]	Compound[b]	Descriptor	RI (SP-1)	OSV[c]
13	geranial	floral/citrus	1244	100.0
5	linalool	lavender	1085	45.2
2	β-pinene/myrcene	piney	984	36.0
11	neral	floral/fresh	1215	34.5
21[e]	decyl acetate	orange	1397	22.6
10	decanal	citrus/waxy	1189	21.9
18[e]	citronellyl acetate	berry/fragrant	1338	20.7
29	hexadecanal	cardboard	1796	19.3
15[e]	(E,E)-deca-2,4-dienal	waxy/citrus	1288	18.3[d]
15	undecanal	green/waxy	1292	18.3[d]
12	geraniol	floral	1230	15.5
14[e]	unknown	fresh floral	1265	13.6
8	terpinen-4-ol	pine/musty	1162	12.7
16[e]	unknown	pine/musty	1299	12.5
27[e]	7-methoxycoumarin	balsamic	1650	12.2
28[e]	unknown	spicy	1674	11.0
4	limonene/cineole	mint	1025	10.7
19	neryl acetate	fruity/fragrant	1342	10.7
24[e]	unknown	spicy	1572	10.6
25	tetradecanal	waxy/floral	1596	10.6
1[e]	(E)-2-hexenal	green	837	10.5
23	germacrene-B	woody/spicy	1544	10.3
7	borneol	musty	1147	9.9
26[e]	unknown	sawdust	1642	9.7
22[e]	unknown	floral	1512	9.6
17[e]	unknown	woody	1310	8.1
6[e]	(E)-2-nonenal	waxy	1132	8.0[d]
6	citronellal	fruity	1134	8.0[d]
20	dodecanal	waxy	1387	7.0
9	α-terpineol	floral	1171	5.9
3	octanal	citrus/soapy	988	4.7

[a] Numbers refer to Figure 1. [b] Compounds identified by retention index, odor descriptor and GC-MS. [c] Odor spectrum value (OSV) is the normalized Charm value (22) adjusted by using Stephen's law with n = 0.5. [d] Compounds coelute. [e] Not previously reported in cold-pressed Key lime oil.

Table II. Most potent odorants found in distilled Key lime oil

Peak No.[a]	Compound[b]	Descriptor	RI (SP-1)	OSV[c]
20	geranial	floral/citrus	1242	100.0
9	linalool	lavender	1085	86.6
18	neral	floral/fresh	1215	19.8
24[e]	unknown	pine/fragrant	1414	19.8
6	limonene/cineole	mint	1025	11.0
28[e]	unknown	floral/spicy	1674	9.6
17	decanal	waxy/citrus	1188	9.5
14[e]	unknown	dandelion	1152	9.2
2[e]	(Z)-3-hexenol	leaf	845	9.0
15	terpinen-4-ol	pine/musty	1160	8.3
19	geraniol	floral	1233	8.2
21[e]	(E,E)-deca-2,4-dienal	waxy/citrus	1288	8.1
22[e]	unknown	berry/woody	1310	8.1
3	2,6,6-trimethyl-2-ethenyl-tetrahydropyran	dandelion	960	6.9
26[e]	unknown	spicy/woody	1573	6.7
1[e]	(E)-2-hexenal	green	837	6.4
13	borneol	earthy	1147	6.4
16	α-terpineol	pine/sweet	1171	6.3
29[e]	hexadecanal	cardboard	1761	6.2
7	2,2-dimethyl-5-(1-methyl-1-propenyl)-tetrahydrofuran	spicy	1035	6.1
10	α-fenchol	pine oil	1094	6.1
12	(Z)-limonene oxide	floral	1127	6.1[d]
12	β-terpineol	musty	1127	6.1[d]
27[e]	unknown	spicy/woody	1658	5.7
25	germacrene-B	woody/musty	1548	5.1
5	1,4-cineole	spicy	1001	4.9
4	dehydrocineole	lemon/mint	975	4.0
11	(E)-limonene oxide	lemon	1120	3.8[d]
11	terpinen-1-ol	musty	1120	3.8[d]
23	dodecanal	waxy	1395	3.2
8	(Z)-furan linalool oxide	floral	1058	2.9

[a] Numbers refer to Figure 2. [b,c,d,e] See footnote in Table I.

Table III. Composition of Distilled Lime Oils (% by weight)

Compound	cold-pressed	hydrodis-tilled, 1 hr	refluxed 6 hr, hydrodistilled	commercial sample
β-pinene	13.1	6.3	0.2	2.7
limonene	43.6	32.6	6.5	29.9
neral	2.8	1.8	~0	1.3*
geranial	4.0	0.9	~0.1	0.4
α-terpineol	1.0	12.5	17.1	10.1
γ-terpineol	~0.1	0.7	6.1	2.5
terpinen-1-ol	~0	0.2	4.5	1.3

*coeluted with significant amounts of carvone, not often found in lime oil (19)

Inc., Geneva, NY) fitted with an SP-1 (Supelco, Inc.) column (15 m x 0.25 mm x 0.25 μm film thickness) and an olfactometer tube which carried the effluent diluted with a stream of humidified air at 40 °C to the nose of the sniffer (22,24). Splitless injection was used and the oven temperature was held at 35 °C for 3 min., then raised at 6 °C/min. to 225 °C and held for 10 min. Retention times of the odorants were recorded and converted to linear retention indices from the retention times of a C_7–C_{18} n-alkane standard by using Charmware™. All dilutions were sniffed twice until no odor was detected in the most dilute sample. A descriptor file was created by sniffing the most concentrated sample several times until a list of the most frequently used terms was acceptable to the sniffer. A modified citrus flavor wheel was used to help develop an acceptable set of odor descriptors (25).

Gas Chromatography-Mass Spectrometry (GC-MS). An HP model 6890 gas chromatograph directly interfaced with an HP model 5985 mass selective detector was fitted with an SP-1 (Supelco, Inc.) column (25 m x 0.25 mm x 0.25 μm film thickness). Mass spectra were recorded in the electron impact mode at 70 eV. The oven temperature program was: 35 °C for 5 min., 4 °C/min to 225, hold for 10 min. Retention indices were recorded for n-alkanes, and used for identifying odorants whose retention indices were found using GCO. Retention indices of known standard compounds were used to identify odorants in the lime oil. The *Registry of Mass Spectral Data*; Wiley, NY was also used for identification of unknown compounds.

Solid phase microextraction (SPME) analysis. The SPME device was purchased from Supelco Inc. (Bellefonte, PA) and the fibers used are shown in Table IV. One drop of lime oil was placed in a 50 mL dark bottle sealed with a cap fitted with a teflon-coated septum. The fiber was conditioned before each extraction by heating it to 225 °C in the injector port of the GC for 5 min. and allowing it to equilibrate at room temperature for 10 min. The oil was stirred at room temperature and the fiber was exposed to the headspace of the lime oil for a measured time (Table V) by piercing the septum. It was then placed in the injector port of the GC operating in splitless mode, and allowed to desorb for 5 min., with the inlet purge flow turned off for 1.0 min. at the time of injection. Absorption by the fiber was measured by recording the FID and total ion chromatograms of the headspace, and by GCO, where odor intensities were recorded for a single run.

Table IV. Types of Fibers Used with Manufacturer's Recommendations.

fiber	phase	film thick-ness (μm)
PDMS	polydimethylsiloxane, nonboned: for volatiles	100
PDMS/DVB	polydimethylsiloxane/divinylbenzene, partially crosslinked: for polar volatiles	65
CAR/PDMS	carboxen™/polydimethylsiloxane, partially crosslinked: for gases and low molecular weight volatiles	75

Table V. Total number of odors detected at different exposure times using GCO

Exposure time (min.)	No. of odors detected
0.5	12
2.0	18
10.0	25

No odors were detected with retention indices higher than 1400 (SP-1).

Concentration of samples by fractionation. Silica gel, activity III, was used to separate 0.5 g of cold-pressed lime oil into fractions by open column chromatography. The column was prepared using hexane and the lime oil was separated into hydrocarbon and oxygen containing fractions. The hydrocarbons were eluted from the column using hexane, then diethyl ether was added in 10% increments to the hexane as described by Dugo et al. (*19*) until all oxygen containing fractions had eluted. The fraction containing aldehydes was analyzed by SPME-GCO and GC-MS.

Results and Discussion

GCO (CHARM) analysis detected between 45-50 odorants in both cold-pressed and distilled Key lime oil samples, and the odor spectrum of the most intense odorants is shown for each in Figures 1 and 2. The odor spectrum values (OSVs) are tabulated in order in Tables I and II, together with additional data to aid in the identification of the odorants. The OSVs are independent of the concentration of the odorant in the sample (*22*), so they provide an estimate of the relative importance of each odorant

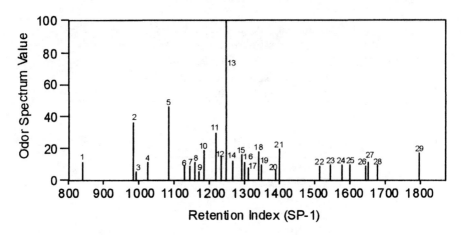

Figure 1. Odor Spectrum of cold-pressed Key lime oil.

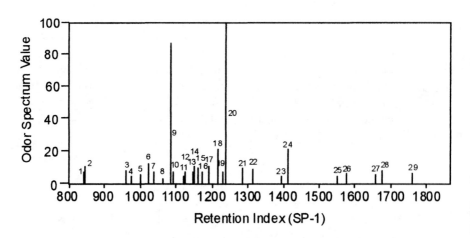

Figure 2. Odor Spectrum of distilled Key lime oil.

in the sample. The aroma of both lime oils is dominated by geranial, linalool and neral, with many additional simple and terpene aldehydes also contributing to the overall aroma. Most of the C_{10} terpene hydrocarbons are known have low odor-activity, but limonene is present in large enough amounts in lime oil to be detected by GCO. β-Pinene was detected in only the cold-pressed oil. Germacrene-B is the only sesquiterpene that was identified, in agreement with earlier work (26) although many odor-active compounds were detected in the region R = 1300–1650, where most sesquiterpenes elute. The compounds had odors described as woody, spicy, sawdust. 7-Methoxycoumarin was identified as a new odorant in cold-pressed lime oil, not previously reported, and was found to have a balsamic, woody tone.

Distilled lime oil has a different chemical composition and aroma from the cold-pressed oil as a result of changes in its composition caused by many acid catalyzed reactions on the cold-pressed oil upon distillation (27). The composition of distilled lime oil can vary widely (Table III), but in all cases much of the monoterpene component is lost to hydration to give a variety of ethers, oxides and alcohols. Many of these such as fenchol, borneol, and β-terpineol have musty odors, and together with α-terpineol are regarded as off-odors in citrus products. These compounds can be seen in Table II and Figure 2 which shows that none of them have large OSVs but many of them play a role in changing the aroma of distilled lime oil compared with that of the cold-pressed oil. More piney, less fragrant were phrases used to describe distilled lime oil. The percentage of α-terpineol can be as high as 20%, and its odor has been described as floral, lilac, but as its detection threshold is higher than other terpene alcohols, its impact is lower.

Tables I and II show many compounds which have not previously been reported as components of lime oil (17-20). SPME was used to trap and concentrate unidentified volatiles and was used in this determination primarily as a screening tool. Figure 3 shows the differences in absorption for selected odorants when different fibers were used. Fiber thickness also played a role, as the thickest fiber consistently gave the highest response from the MSD. Volatiles such as limonene tended to overload the fiber under most operating conditions because it is present in much higher concentrations than any other lime oil component. The PDMS/DVB fiber was less sensitive to oxygen containing compounds than the other fibers examined, and it was found to be of least value in this application because many of the unidentified odorants were believed to contain oxygen. The PDMS fiber was the most sensitive, and with its broad range of selectivity and superior sensitivity, it was the fiber of choice, in agreement with other SPME applications to flavor analysis (11,13,28).

The effect of varying the time of exposure was examined with the goal of determining the optimum time for detecting the largest number of different odorants by GCO. Since the time for equilibration varies with the solubility of the analyte in the fiber, and its ability to leave the food matrix, it was necessary to vary the exposure time to detect specific analytes. Figure 4 shows that for many monoterpenes, increasing the exposure time beyond 2 minutes caused little improve-

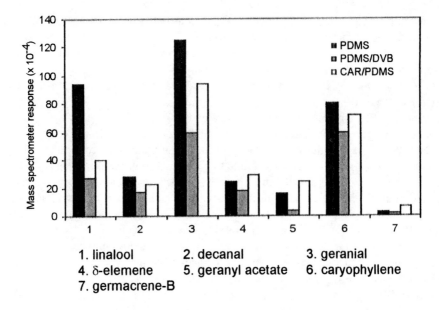

1. linalool   2. decanal   3. geranial
4. δ-elemene   5. geranyl acetate   6. caryophyllene
7. germacrene-B

Figure 3. Effect of fiber type on adsorption measured by mass spectrometer response (area count)

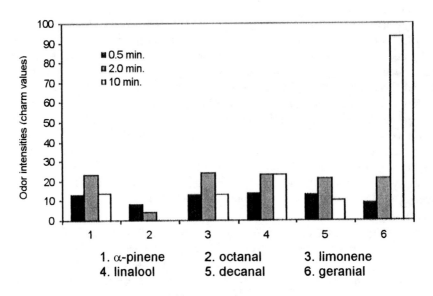

1. α-pinene   2. octanal   3. limonene
4. linalool   5. decanal   6. geranial

Figure 4. Effect of exposure time on adsorption by a PDMS fiber as measured by odor intensities.

ment in GCO detection, with the exception of geranial for which there was close to a linear relationship between exposure time and odor intensity. Deibler et al. found that the odor intensities measured as odor spectrum values obtained from SPME and from serial solvent extraction are not identical (*16*). Table V shows that the number of odorants detected increased with exposure time, but no odorants were detected above RI = 1400 (SP-1) because their volatility was too low for them to be adsorbed by the fiber. The FID chromatograms indicated very little activity above RI = 1400. In a study of the essential oil of hops, Field et al. used exposure times of up to 24 h to detect sesquiterpenes (*28*), and a 90 min. exposure time was used to examine the headspace of apples (*29*). The sesquiterpenes detected in these studies are not odor-active, but we were unable to detect any odor-activity using an exposure time of 30 min., even though germacrene-B had been detected using GCO.

Table VI shows that there are many regions in the chromatogram where some of the most potent odorants in lime oil coelute. An SP-10 column (Supelco, Inc.) was used to resolve some of the coeluting compounds, but it was found to be less effective overall than an SP-1 column for identifying unknown odorants. Many of the monoterpene oxygen containing compounds have retention indices similar to those of the sesquiterpenes on an SP-10 column, which caused coelution in a region where improved resolution was essential for identification of unknown odorants.

Table VI Coeluting odorants found in lime oil (SP-1)

Retention Index (SP-1)	Odorants (in order of elution)
960-2	2,2,6-trimethyl-2-ethenylpyran, β-pinene
985-6	octanal, myrcene
1005-12	1,4-cineole, α-terpinene, *p*-cymene
1017, 1021, 1035*	1,8-cineole, limonene, 2,2,dimethyl-5-(1-methyl-1-propenyl)-tetrahydrofuran
1085-7	nonanal, linalool
1123-4	(E)-limonene oxide, β-terpineol
1146-8	borneol, *p*-methylacetophenone
1131-2	(E)-2-nonenal, citronellal
1216-20**	nerol, citronellol, neral, carvone
1283-4	undecanal, (E,E)-2,4-decadienal

* detection depends upon the amount of limonene present.
** detection depends upon the amount of neral present.

The odorants detected by the use of SPME and GCO on cold-pressed oil extracts concentrated by fractionation are shown in Table VII for the aldehyde fraction. The adsorption of aldehydes increased with the time of exposure and the subsequent identification of aldehydes was improved over that for the whole lime oil. The relative amount of each aldehyde found in the fraction was different from that found in the whole oil, and the compound of highest odor intensity at RI = 2017 (SP-10) was present in greater quantity than was found in the whole oil. The retention index and mass spectrum tentatively identify the compound as caryophyllene oxide, but if this is the case, then it is present in much larger amounts in the fractionated sample than in the original oil. Other compounds detected in larger amounts than expected from the analysis of the whole oil are shown in Table VIII. These results suggest that fractionation on silica gel has caused some distortion of the sample. Concerns about isomerization and formation of artifacts caused by fractionation on silica gel have been reported (*17,30*).

Conclusions

Many of the major odorants in cold-pressed and distilled Key lime oils have been identified, including some new compounds not previously reported in lime oils. Most of the unidentified odorants eluted among the sesquiterpenes and were found to contribute a woody spicy tone to Key lime oil. SPME was found to be a rapid and

Table VII. Compounds found in the aldehyde fraction by using SPME-GCO.

Compound	Descriptor	RI (SP-10)	Charm value
hexanal	green	1092	13
cineole/2-hexenal	eucalyptus	1207	15
unknown	waxy/floral	1249	8
octanal	citrus/green	1297	19
unknown	waxy/floral	1363	20
nonanal	citrus/waxy	1404	14
unknown	mushroom	1416	10
citronellal	fruity	1491	12
decanal	citrus/waxy	1508	20
unknown	spicy	1560	11
citronellyl acetate	berry/fragrant	1692	11
neral	floral/fresh	1706	16
dodecanal	waxy	1726	15
geranial	floral/citrus	1754	13
perilla aldehyde	spicy	1817	18
tridecanal	waxy	1833	9
tetradecanal	waxy/floral	1940	16
C_{15} oxide	sawdust/spicy	2017	23
hexadecanal	woody	2154	10

Table VIII. Compounds found in increased amounts in cold-pressed lime oil after fractionation on silica gel

Compound	RI (SP-10)
(E)- and (Z)-limonene oxide	1467, 1480
epoxyterpinolene	1486
ethyl nerate	1698
ethyl geranate	1762
caryophyllene oxide	2025
humulene oxide	2083
many C_{15} alcohols	>2000

versatile method of sample preparation, and when used in conjunction with GCO as a screening tool, it became possible to examine very large numbers of samples under a variety of conditions from fractionation experiments. The conditions required for equilibration for different odorants varied, so a range of sampling conditions was used to optimize the extraction of each odorant. By using a 100 μm PDMS fiber, it was possible to overload the fiber at lower retention indices in order to maximize the adsorption of sesquiterpenes. Very little odor-activity was detected above RI = 1400 (SP-1 column) or RI = 2230 (SP-10 column) except in the concentrated samples from the fractionation experiments. The oxygen containing samples showed evidence of artifact formation after elution over silica gel, suggesting that identification of these odorants may be difficult, even though many of them possess considerable odor-activity.

References

1. Arthur, C. L.; Killam, L. M.; Buchholz, K. D.; Pawliszyn, J.; Berg, J. R. *Anal. Chem.* **1992**, *64*, 1960-1966.
2. Potter, D. W.; Pawliszyn, J. *J. Chromatogr.* **1992**, *625*, 247-255.
3. Zhang, Z.; Yang, M. J.; Pawliszyn, J. *Anal. Chem* **1994**, *66*, A884-A852.
4. Garcia, D. D. l. C.; Reichenbächer, M.; Danzer, K.; Hurlbeck, C.; Bartzsch, C.; Feller, K.-H. *J. High Resolut. Chromatogr.* **1997**, *20*, 665-668.
5. Yang, X.; Peppard, T. *J. Agric. Food Chem.* **1994**, *42*, 1925-1930.
6. Zhang, Z.; Pawliszyn, J. B. *Anal. Chem.* **1993**, *65*, 1843-1852.
7. Keszler, A.; Héberger, K. *J. High Resolut. Chromatogr.* **1998**, *21*, 368-370.
8. Yang, X.; Peppard, T. *LC-GC* **1995**, *13*, 882-884, 886.
9. Coleman III, W. M. *J. Chromatogr. Sci.* **1997**, *35*, 245-248.
10. Coleman III, W. M.; Lawrence, B. M. *Flavour Fragrance J.* **1997**, *12*, 1-8.

112

11. Miller, M. E.; Stuart, J. D. *Anal. Chem.* **1999**, *71*, 23-27.
12. Steffan, A.; Pawliszyn, J. B. *J. Agric. Food Chem.* **1996**, *44*, 2187-2193.
13. Jia, M.; Zhang, Q. H. *J. Agric. Food Chem.* **1998**, *46*, 2744-2747.
14. Coleman III, W. M.; Perfetti, T. A.; Lawrence, B. M. *J. Chromatogr. Sci.* **1998**, *36*, 575-578.
15. Ong, P. K. C.; Acree, T. E.; Lavin, E. H. *J. Agric. Food Chem.* **1998**, *46*, 611-615.
16. Deibler, K. D.; Acree, T. E.; Lavin, E. H. *J. Agric. Food Chem.* **1999**, *47*, 1616-1618.
17. Chamblee, T. S.; Clark, B. C., Jr. *J. Essent. Oil Res.* **1997**, *9*, 267-274.
18. Dugo, P.; Mondello, L.; Lamonica, G.; Dugo, G. *J. Agric. Food Chem.* **1997**, *45*, 3608-3616.
19. Dugo, P.; Cotroneo, A.; Bonaccorsi, I.; Mondello *Flavour Fragrance J.* **1998**, *13*, 93-97.
20. Shaw, P. E. In *Qualitiy Control Manual in Citrus Processing Plants*; Redd, J. B., Shaw, P. E., Hendrix, C. M., Hendrix, D. L., Eds.; AgScience, Inc.: Auburndale, FL, 1996; Vol. III, pp 134-172.
21. Chisholm, M. G.; Wilson, M. A. In *Tenth National Conference on Undergraduate Research*, Proceedings; Yearout, R. D., Ed.: Salisbury State University, MD; 1998; Vol. V, pp 1751-1755.
22. Acree, T. E. *Anal. Chem.* **1997**, *69*, 170A-175A.
23. Marin, A. B.; Acree, T. E.; Barnard, J. *Chem. Senses* **1988**, *13*, 435-444.
24. Acree, T. E.; Barnard, J.; Cunningham, D. G. *Food Chem.* **1984**, *14*, 273-286.
25. Tamura, H.; Yang, R.-H.; Sugisawa, H. In *Bioactive Volatile Compounds from Plants*; Teranishi, R., Buttery, R. G. Sugisawa, H., Eds.; ACS Symposium Series 523; American Chemical Society: Washington, DC, 1993; pp 121-136.
26. Clark, B. C., Jr.; Chamblee, T. S.; Iacobucci, G. A. *J. Agric. Food Chem* **1987**, *35*, 514-518.
27. Clark, B. C., Jr.; Chamblee, T. S. In *Off-Flavors in Foods and Beverages*; Charalambous, G., Ed.; Dev. Food Sci.; Elsevier Sci. Publ.: Amsterdam, 1992; Vol. 28, pp 229-285.
28. Field, J. A.; Nickerson, G.; James, D. D.; Heider, C. *J. Agric. Food Chem.* **1996**, *44*, 1768-1772.
29. Matich, A. J.; Rowan, D. D.; Banks, N. H. *Anal. Chem.* **1996**, *68*, 4114-4118.
30. Chamblee, T. S.; Clark, B. C., Jr.; Radford, T.; Iacobucci, G. A. *J. Chromatogr.* **1985**, *330*, 141-151.

Chapter 10

SPME Analysis of Flavor Components in the Headspace of Scotch Whiskey and Their Subsequent Correlation with Sensory Perception

John Conner, Kenneth Reid, and Graeme Richardson

Scotch Whiskey Research Institute, The Robertson Trust Building, Research Park North, Edinburgh EH14 4AP, United Kingdom

To understand the development of aromas in mature distilled spirits, a full appraisal of flavour active compounds present in whiskies was conducted using gas chromatography with sniff port and mass spectral detection. More than 90 'aroma zones' could be identified and the majority correlated with peaks / ions from the mass spectral data. Analysis of spirit headspace using solid phase micro-extraction and gas chromatography with multiple ion monitoring mass spectrometry detection was used to determine the relative volatility of 'aroma zone' compounds. The method was used to measure differences in the headspace of a number of product types and to follow changes with maturation quality, thus providing a clearer and more relevant correlation with descriptive sensory terms.

With improved analytical methods there has been a large increase in the number of compounds identified in whisky. This has not resulted in a better understanding of flavour. Conventional methods of analysis working from liquid concentrations fail to provide a good correlation with sensory properties. One reason is the effect of matrix interactions, where the distribution of volatile constituents between the liquid and its headspace is regulated by the concentration of other (frequently non - volatile) components (*1,2*).

Scotch whisky is a matured spirit, with a legal minimum of 3 years ageing in small oak casks. Maturation provides an example of the problems encountered in trying to relate chemical and sensory analysis. During maturation both volatile and non-volatile components are extracted from the cask causing marked sensory changes in the properties of the spirit. The appearance of mature, wood derived aromas accompanies the disappearance of harsh, unpleasant characters present in the new distillate. Levels of wood aromas vary between different whiskies. Strong wood aromas of a new charred cask are required for bourbon, but for Scotch whisky a less prominent wood character is required and cask re-use is normal. Cask re-use greatly reduces the levels of extracted wood components and it is possible to correlate, generally, the levels of these components to the perception of mature characters such as *sweet* and *vanilla* (3). However cask re-use affects the levels of all wood components and so far it has not been possible to identify which individual wood components give rise to mature character in spirits.

Analysis for changes in distillate volatiles as a result of maturation shows quantitative rather than qualitative changes in the composition. The decrease in unpleasant aromas during maturation may result from changes to potent odourants not readily detectable by conventional means. Alternatively changes in the whisky matrix may alter the release of existing components or the presence of wood aromas may mask or alter the perception of distillate components. Identifying important aroma compounds extracted from wood during maturation would be an important step in understanding the mechanisms involved in whisky maturation.

This study describes the investigation of woody aromas in matured grain spirits using gas chromatography - olfactometry and headspace analysis using solid phase micro-extraction (SPME). Gas chromatography – olfactometry (GC/O) was used to characterise and investigate the origin of the most intense aroma zones extracted from cask wood during maturation. SPME headspace analysis was used to calculate odour activity values for identified compounds (4). Past research has shown that other whisky components can suppress the release of hydrophobic aroma compounds (1,2) and that odour activity values calculated from liquid concentrations would accordingly over estimate their activity. Consequently for this experiment odour activity values were calculated by comparison of peak areas from the headspace of whiskies and standard solutions of aroma compounds at their odour threshold concentration.

Experimental

Samples and Sample Preparation

Extracts of wood aromas were prepared by soaking 10 grams of toasted oak shavings in 500 mL of 60 % ethanol for 1 month. Grain whiskies were used for these experiments. Grain whiskies are distilled to a higher ethanol concentration and contain much lower levels of distillate congeners than malt whiskies. Interference between wood and distillate aroma compounds is virtually eliminated by the use of grain spirits. The grain whisky (70 % ethanol v/v) used in this experiment had been matured for 6 and 48 months in new toasted oak casks and standard refill Scotch whisky casks.

Each whisky (20 mL) and wood extract (20 mL) were mixed with 5 mL of dichloromethane and then diluted to 10 % ethanol (v/v) with 20 % NaCl in 0.05M hydrochloric acid. The separated dichloromethane layer was reduced to 500 μL in a stream of nitrogen before analysis.

Gas Chromatography - Olfactometry

GC/O analysis used a TRIO 2 GC-MS fitted with a Gerstell odour port. Effluent from the column was split between the MS and odour port with an approximate split ratio of 5:1 in favour of the odour port. Gas chromatography used a 0.32 mm x 50 m CP-WAX 57CB column (df = 0.5 μm). Initial oven temperature was 40 °C, increasing after 5 minutes to 240 °C at 5 °C min^{-1}. Injector temperature was 240 °C and transfer line to the MS and odour port were at 250 °C. Extracts for GC/O (2 μL) were injected splitless with the split valve opening 30 seconds after injection. Timing and description of aromas were recorded by two assessors from 5 to 50 minutes after injection. Each assessor analysed each extract / dilution in duplicate. For comparisons between extracts, aromas appearing in only 1 run were ignored. For the 48 month samples aroma extract dilution analysis was used to identify the most potent odourants in the extracts (4). Aromas were classified by flavour dilution factor as low intensity (present in original extract), medium intensity (present in 1 in 5 dilution of original extract) or high intensity (present in 1 in 25 dilution). In the dilution analysis only aromas consistently identified by both assessors were included. Calibration of aroma times with chromatogram retention times was initially achieved using a set of odour standards. Within runs this was monitored using 3 compounds with a distinctive aroma present in wood - acetic acid, *cis* oak lactone and vanillin. Variation in retention times for 'aromagrams' and mass chromatograms were identical (standard deviation of approximately 1 second) with aroma detection starting 2 - 4 seconds before the MS retention time at maximum peak height. In this study the assumption is made that an aroma zone represents one aroma compound. However this may not always be the case as overlapping of aroma compounds may result in one or other being missed or masked.

Headspace SPME Analysis

Headspace analysis used a 85 μm polyacrylate SPME fibre. Whiskies were diluted to 20 % ethanol v/v, 5 mL placed in 10 mL headspace vial sealed with a PTFE silicone septum and the fibre exposed for 50 minutes at 37 °C. For SPME analysis the injector temperature was increased to 270 °C and a 0.75 mm ID liner used. The fibre was desorbed for 5 minutes and the column was routed only to the MS. Otherwise column and temperature gradient were the same as for GC/O analysis. Compounds were detected using multiple ion monitoring of prominent ions from the mass spectra (see Table II). Odour activity values (*4*) were calculated by dividing the peak area obtained for the whisky by the peak area from a standard solution at odour threshold concentration. Standard solutions were prepared in 20 % ethanol to match the alcoholic strength used in the sensory and headspace analyses of the whiskies. Each sample / standard solution was analysed in duplicate and additional standard concentrations up to 50 times odour threshold were analysed to ensure that the peak area obtained from a whisky came from the linear portion of the liquid concentration : headspace response curve i.e. analyte had not formed a saturated solution.

Sensory Analysis

Sensory tests were performed by 15 members of the Institute sensory panel. Members of the panel are selected by the ability to recognise and describe aromas and trained in the descriptive sensory analysis of distilled spirits. Triangle tests were used to determine whether differences between maturation ages were significant and to describe the differences detected. Samples were diluted to 20 % ethanol and assessed under red light. Sample presentation was fully randomised - experimental design and data analysis used Compusense v3.0 (Compusense Inc., Ontario, Canada).

Results and Discussion

Gas Chromatography - Olfactometry

A number of comparisons were made to characterise the important aroma compounds extracted from wood during maturation. Initially extracts of toasted oak were investigated to detect aromas extracted directly from the wood and which were unchanged by maturation. Next, whisky from new toasted oak casks and refill casks were compared to find out which aromas persist with cask re-use. Finally, for both cask types, spirit matured for only 6 months was compared with 48 month samples to detect any aromas that are formed or degraded during maturation.

Grain Whisky Matured in New Toasted Oak Casks
The results clearly show that wood aromas are extracted directly from the wood of the cask with little subsequent modification during the maturation period. Table I shows that all but one of the medium and high intensity aromas (present in 1 in 5 and 1 in 25 dilutions respectively) were detected in the laboratory prepared wood extract. A total of 44 aromas were detected from the concentrated extract of the grain whisky matured in new toasted oak casks for 48 months. This is considerably less than the 70 detected in the extract of toasted oak but most probably reflects the different wood to spirit ratios. The amount of wood used for

the rapid preparation of extracts was far higher than encountered in cask maturation, giving more wood components above their odour port detection threshold. Dilution of the whisky extract further supports this explanation with a large reduction on the number of aromas detected, decreasing to 19 with the 1 in 5 dilution and to only 7 with a 1 in 25 dilution.

Comparison of new cask samples showed that no additional aromas were present at 6 months maturation. Maturation does not appear to form or degrade aroma compounds, but rather increases concentrations of aroma compounds due to the prolonged extraction of the wood. All the most intense aromas (present 1 in 25 dilution) and the majority of others in the 48 month samples were present after only six months maturation. This agrees with past research which has shown a rapid extraction of wood components from new casks in the early stages of maturation (5).

Grain Whisky Matured in Refill Casks

Comparison between different cask types suggests two distinct origins for aroma compounds in refill casks: from the wood itself and those that originate from previous fills of the cask. For wood aromas, the majority of medium and high intensity wood components from the new toasted oak casks were still detectable in refill casks. GC/O flavour dilution factors were generally lower, with all odours eliminated by a 1 in 25 dilution. This reflects the depletion of wood aromas that occurs with repeated cask use.

Other aromas unique to refill casks were identified as typical constituents of malt spirit (Table II). The most likely origin of these is adsorption by the cask wood during previous fills with Scotch malt whisky. The GC/O flavour dilution factor of two other aromas (23.*20 stale, sweaty* and 28.37 *floral*) increased in the refill casks. These were identified as 3-methylbutanoic acid and 2-phenylethanol which are also constituents of malt spirit.

For refill casks some additional aromas were noted in the six month samples. These were described as floral, nutty, smoked and burnt. The additional aromas could not be related to any of the wood components encountered so far and are most probably components from previous fills, such as aldehydes or acids, the aroma impact of which are subsequently reduced during maturation by esterification or acetal formation.

Headspace Analysis of Known Compounds

To identify which compounds revealed by the dilution procedures actually contribute to the aroma of the grain whisky, SPME headspace analysis was used to calculate odour activity values. The odour thresholds of ten of the identified compounds in Table II have previously been determined at 20 % ethanol in rectified spirit (6,7). Headspace peak areas from the SPME analysis of diluted whiskies were divided by peak areas obtained from standards at odour threshold concentrations (Table III).

Table I. Aromas in the extracts of new toasted oak and grain whiskies matured in refill and new toasted oak casks for 6 and 48 months.

Retention Time	Descriptor	Refill 6 mth	Refill 48 mth	New, Toasted 6 mth	New, Toasted 48 mth	Wood Extract
10.48	higher alcohol	+[b]	**[d]	n.d.[a]	n.d.	n.d.
11.02	fruity, ester	+	**	n.d.	n.d.	n.d.
16.40	solvent, fruity	n.d.	*[c]	n.d.	**	n.d.
17.53	vinegar	+	*	+	**	+
18.78	waxy, floral	n.d.	*	+	***[e]	+
22.15	rancid	n.d.	*	n.d.	**	+
23.20	stale, sweaty	+	**	+	*	+
24.27	waxy, floral	+	**	+	***	+
26.12	floral, fruity	n.d.	**	n.d.	n.d.	n.d.
26.22	tobacco, leafy	+	**	+	**	+
27.08	vanilla, sweet	n.d.	n.d.	+	**	+
27.90	coconut	n.d.	*	+	**	+
28.37	floral	+	**	n.d.	*	+
29.25	sweet, coconut	+	**	+	***	+
30.67	herbal, minty	+	**	+	**	+
31.23	baked	n.d.	*	+	**	+
31.82	sweaty, rancid	+	**	+	**	+
32.23	herbal, floral	n.d.	*	+	***	+
33.33	cloves	+	**	+	***	+
33.80	floral	n.d.	n.d.	n.d.	**	+
34.60	spicy, curry	+	**	+	***	+
36.77	floral	+	*	+	**	+
40.72	vanilla	+	**	+	***	+
41.95	floral	n.d.	*	+	**	+

a	n.d.	not detected
b	+	detected in extract - no dilution analysis
c	*	low intensity aroma – present in concentrated extract
d	**	medium intensity – present in 1 in 5 dilution of extract
e	***	high intensity – present in 1 in 25 dilution of extract

Table II. Aroma compounds identified in 1 in 5 dilutions of extracts from 48 month grain whisky matured in new toasted and refill casks.

Retention Time	Descriptor	Compound	Identity	Threshold[a]
10.48	*higher alcohol*	3-methyl-1-butanol	ms[b] std[c]	43
11.02	*fruity, ester*	ethyl hexanoate	ms, std	0.2
16.40	*solvent, fruity*	?	-	-
17.53	*vinegar*	acetic acid	ms, std	67
18.78	*waxy, floral*	benzaldehyde	ms	1.7
22.15	*rancid*	butyric acid	ms, std	2
23.20	*stale, sweaty*	3-methylbutanoic acid	ms, std	0.9
24.27	*waxy, floral*	?	-	-
26.12	*floral, fruity*	2-phenylethyl acetate	ms, std	1.5
26.22	*tobacco, leafy*	?	-	-
27.08	*vanilla, sweet*	2-hydroxy-3-methyl cyclopentenone	ms	-
27.90	*coconut*	trans oak lactone	ms, std	-
28.37	*floral*	2-phenylethanol	ms, std	60
29.25	*sweet, coconut*	cis oak lactone	ms, std	0.06
30.67	*herbal, minty*	4-ethyl-2-methoxyphenol	ms, std	0.09
31.23	*baked*	?	-	-
31.82	*sweaty, rancid*	octanoic acid	ms, std	3
32.23	*herbal, floral*	4-propyl-2-methoxyphenol	ms	-
33.33	*cloves*	eugenol	ms, std	0.18
33.80	*floral*	?	-	-
34.60	*spicy, curry*	?	-	-
36.77	*floral*	?	-	-
40.72	*vanilla*	vanillin	ms, std	0.1
41.95	*floral*	?	-	-

a   mgL^{-1} in 23 % ethanol from references 6 and 7
b   identified by mass spectrum
c   identity / aroma character confirmed by comparison with standard.

After 4 years maturation only ethyl hexanoate, cis oak lactone and vanillin are present in the headspace above their odour threshold concentration. Two others, 3-methyl-1-butanol (refill casks) and eugenol (new casks), were close to their odour threshold (odour activity values of 0.8 - 0.9). The remaining compounds are all well below their odour threshold. Consequently for distilled spirits, detection of an aroma compound by GC olfactometry does not guarantee that it will produce a sensory response in the sample.

Table III shows that wood components are depleted at different rates with cask reuse. In refill casks after 48 months maturation, the odour activity value for vanillin was approximately half that in a new toasted oak cask. For oak lactone, the rate of depletion was much greater with just over a quarter of the odour activity of the new toasted cask surviving to the refill cask. The different balance of wood aromas in refill cask spirits would produce a different sensory response. Odour activity values generally increased between 6 and 48 months except for 'previous use' components in new casks and benzaldehyde, which only marginally increased in both cask types.

Table III. Quantifying ions and aroma activity values for aroma compounds in spirit matured for 6 and 48 months in new and refill casks. Sample were diluted to 20 % ethanol for analysis

Compound	m/z	Refill (6 mth)	Refill (48 mth)	New (6 mth)	New (48 mth)
3-methy-1-butanol	70	0.44	0.9	0.005	0.006
ethyl hexanoate	88	0.1	1.7	0.09	0.65
benzaldehyde	106	0.004	0.005	0.004	0.005
2-phenylethyl acetate	104	0.003	0.003	0.002	0.001
phenylethanol	91	0.008	0.019	0.001	0.0007
cis oak lactone	99	0.94	2.9	6.2	20.1
4-ethyl-2-methoxyphenol	137	0.22	0.43	0.1	0.09
eugenol	164	0.14	0.2	0.5	0.8
vanillin	151	2.4	5.1	4.2	10.9

Odour activity values of acids were not calculated. Solution - headspace distribution curves were exponential because only unionised molecules partition into headspace. The ratio of unionised to ionised molecules increases with acid concentration (decreasing solution pH) and proportionally more acid is available to partition into the headspace. Consequently partitioning between the solution and headspace is a function of both pH and concentration. Buffering pH to that of mature whisky has so far failed to produce a satisfactory model for volatile release (*8*), perhaps because interactions between organic acids and the aggregation of ethanol may occur that affect headspace partitioning (*9*).

The 6 and 48 month samples for both cask types were assessed by the Institute's sensory panel. Both six month samples were generally described as bland. For refill casks, 7 out of 15 panellists correctly separated the different ages (not significant, $p > 0.05$). For new casks, 12 out of 15 panellists correctly separated the different ages, significant at the 0.1% level. Generally increased *sweet, vanilla* character was noted but additional descriptors were also applied to the 48 month samples such as *burnt wood, resinous* and *earthy*.

The sensory assessment generally agrees with the odour activity values. For refill casks aroma compounds are present at low odour activity values (Table III) and consequently do not have a large impact on whisky aroma. Even after 48 months maturation the spirit was described as bland. This discription was also applied to the 6 month spirit from the new casks. However for these casks the increase in concentrations with time is such that both oak lactone and vanillin approach their recognition thresholds (*7*) and make a much more noticeable impact on the aroma of the spirit. GC olfactometry descriptors for these two compounds (Table II) show that they would be responsible for *sweet, coconut* and *vanilla* aromas in the spirit. However other descriptors such as *burnt wood, resinous* and *earthy* were also used to describe the spirit and these may relate to other aromas in Table I which have not been characterised.

This work represents one small piece in the jigsaw that makes up whisky flavour. Much more work is needed to develop our understanding of the relationship between analytical measurements and the sensory properties of a spirit. This work used grain spirit where little or no interactions with distillate aromas occur. Therefore it may not be directly applicable to malt spirits, where the greater potential for distillate aroma interactions may give a different relationship between odour activity values (calculated from headspace) and sensory response. Developing this relationship will require detailed knowledge of the aroma compounds present in the headspace, their predicted sensory impact and their actual sensory impact. The combination of GC Olfactometry and SPME headspace analysis will be an indispensable tool in providing this information.

Conclusions

This work illustrates the benefit of combining GC Olfactometry with a headspace technique such as solid phase microextraction. It shows that for distilled spirits, not every aroma zone identified by GC Olfactometry will have a significant sensory impact. SPME headspace analysis avoids the problems of matrix interactions and can be used to determine odour activity values for aroma compounds with defined sensory properties in aqueous ethanol. Where such data is available it is possible to convert headspace measurements into a form that can be directly related to the response of a sensory panel.

Using this method it was shown that, from 10 identified aroma compounds only 2, cis oak lactone and vanillin, were present in the headspace of mature whisky above their odour threshold. The levels of these two components measured in whisky headspace illustrates the different impact of wood aromas for spirits matured in new heat treated casks (Bourbon) and refill casks (Scotch Whisky). For whisky matured in new casks the level of these compounds in the headspace is close to their recognition threshold and this gives a noticeable *sweet vanilla* aroma to the mature product. When matured in refill casks the level in headspace is close to their detection threshold and the compounds do not have a marked affect on spirit aroma.

References

1. Conner, J.M.; Paterson, A. and Piggott, J.R. *J. Agric. Food Chem.* **1994**, 42, 2231 – 2234.
2. Conner, J.M.; Paterson, A. and Piggott, J.R. *J. Sci. Food Agric.* **1999**, *79*, 1015 – 1020.
3. Piggott, J.R.; Conner, J.M.; Paterson, A. and Clyne, J. *Int. J. Food Sci. Technol.* **1993**, 28, 303 – 318.
4. Grosch, W. *Flavour and Fragrance Journal*, **1994**, 9, 147 – 158.
5. Liebmenn, A.J. and Scherl, B. *Ind. Engineer. Chem.* **1949**, 41, 534 – 539.
6. Perry, D.R. In *Distilled Beverage Flavour: Recent Developments*, Piggott, J.R. and Paterson, A., Eds., Ellis Horwood, Chichester, UK, 1989, pp 200 – 207.
7. Lee, K.Y.M.; Paterson, A.; Piggott, J.R. and Richardson, G.D. In *Proceedings of the Fifth Aviemore Conference on Malting, Brewing & Distilling*; Campbell, I., Ed.; Institute of Brewing: London, U.K, 1999; pp 335 – 338.
8. Conner, J.M.; Paterson, A. and Piggott, J.R. *J. Sci. Food Agric.* **1994**, 66, 45 –53.
9. Conner, J.M.; Paterson, A. Birkmyre, L. and Piggott, J.R. *J.Inst.* Brew. **1999**, 105, 287 – 291.

Chapter 11

Solid-Phase Microextraction Combined with Gas Chromatography–Olfactometry for Analysis of Cheese Aroma

Jean-Pierre Dufour, P. Delbecq, and L. Perez Albela

Food Science Department, University of Otago, P.O. Box 56, Dunedin, New Zealand

Solid-phase microextraction (SPME) has been shown to be a simple and effective tool for detecting trace levels of flavor compounds in foods and beverages. The aim of this study was to evaluate the application of SPME combined with gas chromatography-olfactometry (GC-O) for the analysis of odorants in Cheddar cheeses. The cheese volatile compounds were extracted by immersing the fiber coating in the sample headspace. Effect of exposure time, influence of absorption temperature (below and above cheese melting point), and fiber coating were investigated. SPME-GCO aroma profiles obtained with several coatings were distinctly different. A total of 45 odor active compounds were detected, the largest odor spectrum being obtained with PDMS-DVB (38 odors) and CAR-PDMS (35 odors).

Cheddar cheese, one of the most popular cheeses worldwide, is characterised by a complex flavor which originates from biochemical and chemical reactions during processing and ripening (1). Despite the significant research done over the past 50 years to characterise the flavor profile and the mode of formation of cheese flavor volatiles, little is known about the odorants determinant of Cheddar cheese aroma (2-6).

Characterisation of the flavor volatiles involves the isolation, separation and identification of the volatile compounds. Methods used to sample Cheddar cheese volatiles include steam distillation (7,8), molecular distillation (9-14), headspace analysis (6,15-19), solvent extraction (6,8,20,21), dialysis (8,22) and direct injection of Cheddar cheese oil obtained by centrifugation (23).

All these methods suffer from numerous drawbacks such as the use of organic solvents, limited sensitivity for low volatility compounds, losses of analytes during the concentration step, large amount of sample required, potential risk of error due to their multistage character, and the necessity for complex apparatus. The use of Solid Phase Micro Extraction (SPME) for flavor analysis can reduce the limitations associated with current methodologies.

SPME is a new solventless isolation method that can be used to extract and concentrate a wide range of volatile compounds from various matrices in a single step (24,25). This technique was initially developed for sampling organic contaminants in water (26), but it has also been applied with success to the analysis of volatile flavor compounds in food and beverages. SPME has the potential to reduce the time required for sampling, works well in combination with any separation and detection systems, has a relative low cost, and only requires a small amount of sample (24,26).

The suitability of SPME for studying and characterizing cheese volatile profiles has been evaluated by Chin et al. (27). Major volatile compounds (fatty acids, delta-lactones) were readily extracted by SPME polydimethylsiloxane and polyacrylate coatings but minor compounds such as sulfur compounds were not observed. Accumulation of these compounds was examined in full- and reduced-fat Cheddar cheese under different ripening conditions (28).

This chapter investigates the use of SPME for the analysis of Cheddar cheese odorants. The extraction efficiency of cheese volatiles with different coatings, the selectivity of these coatings towards a series of homologues of volatile compounds present in cheese and the effect of sampling temperature on the extraction of Cheddar cheese flavor volatiles were analysed.

Materials and Methods

Cheese Samples

Twelve blocks of Cheddar cheese (1 kg loaves) processed under identical conditions and ripened for 12 months were donated by Mainland Cheese Factory (Dunedin, New Zealand).

Sample Preparation

Upon arrival, the cheese loaves were cut into approximately 2 cm by 2 cm pieces, frozen in liquid nitrogen in a Dewar flask and powdered using a Waring blender (Watson Victor Ltd, Dunedin, New Zealand). The powdered cheese was kept in 120 ml capped plastic containers (Tecpak, Dunedin, New Zealand) with no headspace at − 80°C until analysis.

Cheese Volatile Sampling Conditions

Ten grams of the powdered frozen cheese were placed in a 20 ml headspace vial (Alltech, Deerfield, IL, USA) covered with a 20 mm chlorobutyl septum (Alltech) and capped with a 10 mm center hole-20 mm diameter aluminium seal (Alltech). A SPME manual holder (Supelco, Bellafonte, PA, USA) equipped with 100 μm polydimethylsiloxane (PDMS), or 65 μm polydimethylsiloxane/divinylbenzene (PDMS-DVB), or 85 μm polyacrylate (PA), or 75 μm carboxen/polydimethylsiloxane (CAR-PDMS), or 65 μm carbowax/divinylbenzene (CW-DVB) SPME fiber coating (Supelco) was used to sample cheese volatiles.

Prior to the analysis, the sample-containing vials were held at 20°C for 1 h. Before use, the SPME fiber coatings were cleaned in a GC injector port at 220 C for 5 min. Immediately afterwards they were inserted into the vial through the septum at 1cm depth. The fiber coating was exposed to the headspace at 20°C in a controlled-temperature room for 1, 3, 6, 9 and 12 h or for 1 h at 60°C using a heating module (Pierce, Rockford, IL, USA).

Gas Chromatography Analysis

GC-FID Analysis

A Fisons 8000 series GC (Carlo Erba Instruments, Milan, Italy) equipped with a split/splitless injector at 220°C and a flame ionisation detector (FID) at 220°C was used. A 1.0 mm i.d. injector glass liner was used to minimise peak broadening. Separation was carried out on a Nukol (Carbowax-20M analog) column (30 m, 0.32 mm i.d., 0.25 μm film thickness; Supelco, Bellafonte, PA, U.S.) using helium as carrier gas (1.4 ml/min, 24 cm/sec, constant flow mode). Detector gas flow rates were; air 280 ml/min (100 kPa) and hydrogen 25 ml/min (50 kPa). The GC oven was initially held at 30°C for 8 min then increased to 190°C at 5°C/min and held at this temperature for 20 min.

Thermal desorption of volatiles absorbed into the SPME fiber coating was carried out in the injector in a splitless mode, where the fiber coating was inserted to a depth of 2.5 cm. The split valve was opened 2 min after injection (split ratio, 35:1). The SPME fiber was held in the injector port for 5 min. Data integration and computation were performed using Chromcard software (Finnigan Instruments, Manchester, UK). Analyses were done in triplicates.

GC-MS Analysis

GC-MS analysis was carried out using a Fisons 8000 Top GC (Carlo Erba Instruments, Milan, Italy) coupled to a Finnigan MAT MD 1000 mass detector (Finnigan Instruments, Manchester, UK). Separation was achieved using a Nukol column (60 m, 0.32 mm i.d., 0.25 μm film thickness; Supelco, Bellafonte, PA, USA) using helium at 1.4 ml/min (30 cm/sec, constant flow mode) as the carrier gas. Injection (220°C) was done in the splitless mode. After 2 min, the split valve was opened (split ratio, 20:1) and the fiber was removed after 5 min. The oven temperature was held at 30°C for 16 min then increased to 190°C at 5°C/min and held at this temperature for 30 min. The mass spectrometer was operated in the electron impact ionisation mode (70 eV). Source and interface temperatures were 195°C and 240°C, respectively. Detector voltage was 300 V; mass range was from 35 to 350 amu; scan rate was 0.9 scan/sec.

GC-O Analysis

GC-O was carried out using a HP5790 series II GC (Hewlet Packard, Avondale, PA, USA). The GC analytical conditions were as described for the GC-FID analysis. The split/splitless injector was equipped with a 0.7 mm i.d. liner. Carrier gas flow was 1.4 ml/min (24 cm/sec). Detector gases were ; air at 250 ml/min and nitrogen (make-up gas) at 50 ml/min. The total effluent from the column was sniffed at a sniff port (Datu Technology, Geneva, USA) held at 220°C. The effluent from the column was mixed with moistened air (50 SCFH) at 50°C. A computer data-handling system (Charmware software, Datu Technology) recorded the response time interval. A series of paraffins was run in the GC-FID mode to establish the retention indices, using a 4 mm i.d. Focusliner (SGE, Melbourne, Australia) as injector liner (split mode; split ratio, 50:1).

Compound identification

Compounds were identified by matching mass spectral data with the Wiley and NIST library of standard compounds. Finnigan Masslab software version 1.4 (Finnigan Instruments, Manchester, UK) was used for data analysis. In order to obtain retention indices, 0.5 μl of a series of paraffins from 6 to 24 carbon atoms at a concentration of 10 mg/ml cyclohexane was injected under the same GC conditions using a 5 mm i.d. liner (split mode; split ratio, 100:1). Identification of compounds was further confirmed by injecting 0.5 μl of authentic compounds at a concentration of 10 mg/ml in ethanol under identical GC conditions (split ratio, 100:1).

Results and Discussion

Effect of Fiber Coating on Extraction of Cheddar Cheese Volatiles at 20°C

The extraction of cheese volatiles into the different fiber coatings was followed at 20 C over 12 h. All fiber coatings exhibited increasing volatile extraction (expressed as the total area counts of all detected absorbed volatiles) over 12 h (Figure 1) with CAR-PDMS extracting by far the largest amount of cheese volatiles, followed in decreasing order by PDMS-DVB, CW-DVB, PA and PDMS. Although the graph of the total area counts suggests that absorption reached equilibrium after 9 to 12 h incubation, some of the individual volatiles may still be increasing, e.g. decanoic acid. The sudden increase in total area counts observed for the PDMS-DVB between 9 and 12 h was attributable to the replacement of the fiber. This data stresses the importance of keeping a detailed record history for each fiber, including fiber cleaning parameters and injection conditions. Each fiber should be clearly identified. Ideally a set of standards should be used to evaluate the consistency of the fiber extraction performance. A sample could also be used as standard and be re-analysed with each batch of samples of subsequent experiments. Variations in standard peak areas could be used to calculate correction factors (28).

As expected the selectivity (number of peaks) also varied with the nature of the fiber coating used (Table I). Extracted cheese volatiles included components such as volatile fatty acids, lactones, sulfur compounds, carbonyl compounds, and esters (data not shown). The bi-polar coating PDMS-DVB, and to a lesser extent CAR-PDMS, showed the largest selectivity (highest number of peaks absorbed). Single polarity coatings (PA, PDMS, CW-DVB) extracted a similar number of peaks (approx. 30 peaks). These findings are in agreement with Clark and Bunch (29), who reported that bi-polar coatings are able to extract a wider variety of compounds than single polarity coatings.

Table I. Number of Compounds Extracted at 20°C into the Fiber Coatings

Fiber Coating	Absortion Time (hours)				
	1	3	6	9	12
PA	20	23	31	31	31
PDMS	17	24	26	30	34
CW-DVB	21	22	28	33	35
PDMS-DVB	37	51	64	81	95[a]
CAR-PDMS	42	46	60	62	62

[a]- A new PDMS-DVB fiber was used for this experimental point

The polarity, the volatility and size of the analytes are among the factors that influence the coating absorption efficiency: e.g. polar analytes are attracted to polar phases and small molecules are better extracted using porous coatings (e.g. Carboxen or DVB coating). A typical example is the extraction of short chain fatty acids. CAR-PDMS coating was very efficient in trapping C-2 to C-4 fatty acids, but absorption efficiency rapidly decreased with increasing chain length (data not shown). Figure 2 illustrates the absorption-time profile of acetic acid using different fiber coatings at 20°C. Unlike CAR-PDMS, the absorption efficiency of the other coatings increased with increasing chain length, the absorption of C-10 acid being very similar for PA, PDMS, PDMS-DVB and CW-DVB coatings (Figure 3).

When analysing the extraction efficiency of selected volatiles (C-2 to C-12 fatty acids, C-3 to C-9 methylketones, diacetyl, acetoin, 2,3-butanediol, ethanol, delta-dodecalactone), CAR-PDMS showed the best extraction performance for low molecular weight compounds whereas PDMS-DVB gave the best results for mid and high molecular weight analytes (data not shown).

For larger molecules such as decanoic acid, the absorption of which increased linearly with time, it is essential to control the sampling conditions in order to get reproducible data. Table II summarises the absorption efficiency (area counts) of the short chain fatty acids and the corresponding coefficient of variation using PA, CAR-PDMS and PDMS-DVB coatings at 20°C.

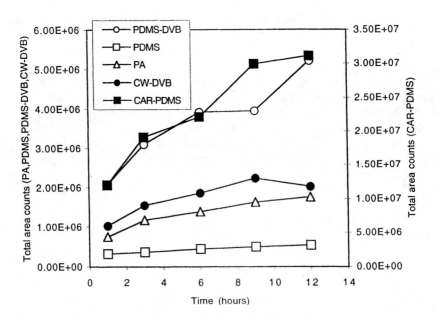

Figure 1. Absorption-time profiles for total cheese volatiles (total area counts) using different fiber coatings at 20°C.

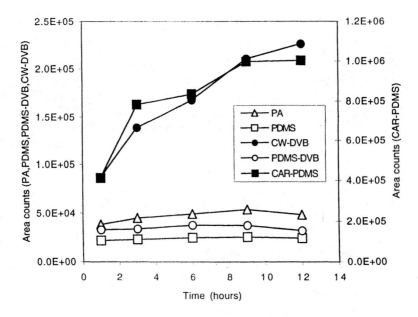

Figure 2. Absorption-time profiles for acetic acid using different fiber coatings at 20°C.

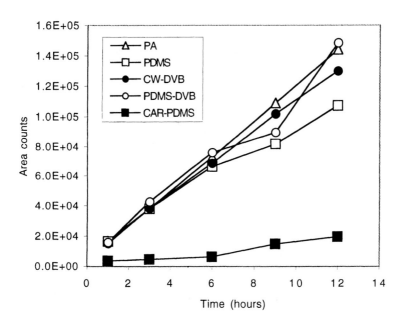

Figure 3. Absorption-time profiles for decanoic acid using different fiber coatings at 20°C.

Table II. Absorption Efficiency of Short Chain Fatty Acids using PA, CAR-PDMS and PDMS-DVB Fibers at 20°C (12 h)

| Acid | Area counts (CV) | | |
	PA	CAR-PDMS	PDMS-DVB
Acetic	48,000 (22.3)	1,002,016 (10.2)	31,891 (14.1)
Butyric	456,375 (5.6)	8,263,821 (7.5)	1,058,323 (18.2)
Hexanoic	542,596 (4.2)	1,269,783 (10.7)	1,819,173 (10.7)
Octanoic	265,103 (1.6)	88,839 (23.6)	351,906 (6.4)
Decanoic	144,904 (3.6)	19,521 (30.4)	148,484 (5.3)

CV: coefficient of variation, in % (n=3)

Effect of Sampling Temperature on Extraction of Cheddar Cheese Volatiles

Sampling temperature is very important for headspace analysis. Higher temperatures will favour higher sensitivity and extraction rate. The increase of temperature will also favour higher extraction yield. A too high temperature will, however, negatively affect the absorption of volatiles, the absorption being an exothermic process. Another critical factor to consider when analysing cheese volatiles by headspace sampling is the effect of increasing temperature on cheese physical state. At 20°C, there was no apparent change in the cheese physical state but as temperature was increased an oily layer exuded from the cheese and covered the surface of the sample.

Absorption of cheese volatiles at 20°C (no change in cheese physical state) was compared to absorption at 60°C (presence of an oily layer). Table III summarises the data obtained for the extraction of short chain fatty acids. As expected, increasing the temperature increased the extraction efficiency, especially of the less volatile compounds as illustrated by the large increase of the 60°C-area counts/20°C-area counts ratio (Table III). Extending the incubation time at 20°C from 1 h to 12 h gave area counts similar to 1 h incubation at 60°C as indicated by the 60°C-area counts/20°C-area counts ratio which ranged from 0.4 to 3 (Table III).

Table III. Effect of Temperature/Time on the Relative Absorption Efficiency of Short Chain Fatty Acids using CAR-PDMS and PDMS-DVB Fibers

| Acid | 60°C-area counts/20°C-area counts ratio | | | |
| | 1 h,60°C versus 1 h,20°C | | 1 h,60°C versus 12 h,20°C | |
	PDMS-DVB	CAR-PDMS	PDMS-DVB	CAR-PDMS
Acetic	1.54	2.90	1.57	1.19
Butyric	0.57	6.00	0.42	1.07
Hexanoic	5.95	13.10	0.73	1.74
Octanoic	15.10	7.01	1.52	2.43
Decanoic	25.40	16.00	2.71	2.99

For compounds such as acetoin and 2-heptanone, incubation at 20°C (12 h) gave overall better extraction yield than at 60°C (1 h), especially when using the PDMS-DVB coating (Table IV).

Table IV. Effect of Temperature/Time on the Relative Absorption Efficiency of 2-Heptanone and Acetoin using CAR-PDMS and PDMS-DVB Fibers

Compound	60°C-area counts/20°C-area counts ratio			
	1 h,60°C versus 1 h,20°C		1 h,60°C versus 12 h,20°C	
	PDMS-DVB	CAR-PDMS	PDMS-DVB	CAR-PDMS
2-Heptanone	0.54	14.30	0.23	0.81
Acetoin	0.30	2.72	0.40	0.58

Figures 4 and 5 illustrate typical chromatograms for cheese volatiles extracted at 20°C (12 h) and 60°C (1 h), using CAR-PDMS and PDMS-DVB coatings, respectively. CAR-PDMS extracted the highly volatile compounds (up to RI of approx. 1,800 on a Nukol column) (RT of 32 min) more efficiently while PDMS-DVB was more efficient for compounds with higher RI. Significant differences were observed between the profiles of cheese volatiles extracted at 20°C (12 h) and 60°C (1 h) using the CAR-PDMS fiber, with more abundant (in yield and number) highly volatile compounds (RT up to 30 min) observed at low temperature (below cheese melting temperature) (Figure 4). A similar trend was observed with the PDMS-DVB fiber, which gave a better extraction yield of the more volatile compounds. The two profiles, however, showed a similar number of peaks (Figure 5).

Analysis of Cheddar Cheese Odorants

Table V summarises the GC-O data of triplicate analyses of cheese volatiles extracted at 20°C (12 h) and at 60°C (1 h) using the PDMS-DVB fiber. Overall, the data were very consistent.

Moreover, the odor profiles for the two sets of conditions were similar. A few aroma-active compounds were only detected when incubation was done at 20 C (12 h) (RI 773-778, sulfurous; RI 1361-1373, fatty acid; RI 1462-1466, aldehydic; RI 1708-1723, broth; RI 1818-1822, mouldy) and at 60°C (1 h)(RI 816-830, solvent; RI 2270-2277, woody; RI 2287-2296, sulfurous), respectively (Table V). The origin of these differences must await further characterisation of the odors. Possible causes include the heat induced release of an oily layer which could act as a barrier for polar molecules but at the same time contribute to the concentration of hydrophobic volatiles. Heat treatment could also destroy thermally sensitive odorants as well as generate new volatiles. The importance of the difference in volatile composition to the characteristic Cheddar cheese flavor remains to be determined.

The five coatings were evaluated under identical conditions for their efficiency in extracting the odor-active compounds of Cheddar cheese at 20°C (12h).

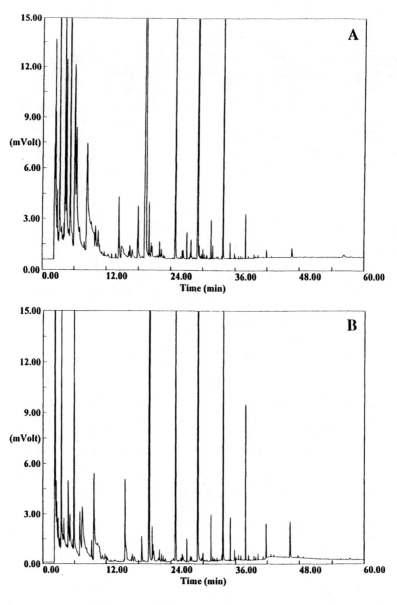

Figure 4. GC-FID profile of cheese volatiles extracted at 20°C (12 h) (A) or at 60°C (1 h) (B) using CAR-PDMS.

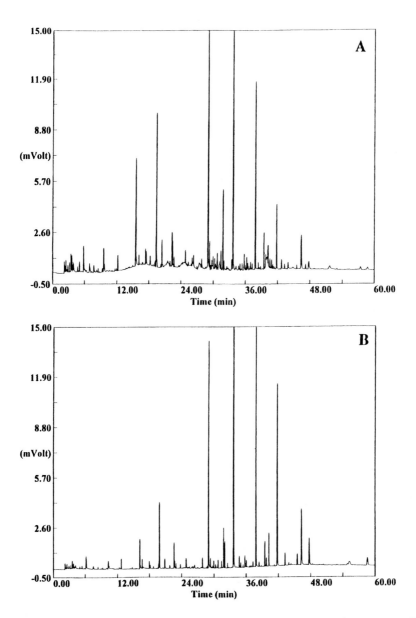

Figure 5. GC-FID profile of cheese volatiles extracted at 20°C (12 h) (A) or at 60°C (1 h) (B) using PDMS-DVB coating.

Table V. Aroma Profiles of Extracted Cheddar Cheese Volatiles using PDMS-DVB Fiber (20°C, 12 h and 60°C, 1 h)

RI	Descriptor	20°C, 12 h			60°C, 1 h		
		1	2	3	1	2	3
753-775	sulfurous	+	+	+	+	+	+
773-778	sulfurous	+	+	+	nd	nd	nd
781-790	sulfurous	+	+	nd	+	+	+
816-830	solvent	nd	nd	nd	nd	+	+
913-929	solvent	+	+	+	+	+	+
1043-1048	fruity	+	+	+	+	+	+
1138-1142	fruity	nd	+	+	nd	nd	+
1240-1250	fruity	+	+	+	nd	+	+
1295-1304	floral	+	+	+	nd	nd	+
1309-1320	mushroom	+	+	+	+	+	+
1361-1373	fatty acid	+	+	+	nd	nd	nd
1385-1391	sulfurous	+	+	+	+	+	+
1411-1422	mouldy	+	+	+	+	+	+
1447-1458	fatty acid	+	+	+	+	+	+
1462-1466	aldehydic	+	+	+	nd	nd	nd
1469-1480	broth	+	+	+	nd	+	+
1518-1524	aldehydic	+	+	+	+	+	+
1548-1558	aldehydic	+	+	+	+	+	+
1600-1610	aldehydic	+	+	+	+	+	+
1635-1662	fatty acid	+	+	+	+	+	+
1665-1670	floral	+	+	+	+	nd	+
1686-1695	fatty acid	+	+	+	+	+	nd
1708-1723	broth	+	+	+	nd	nd	nd
1720-1732	aldehydic	+	+	+	nd	nd	+
1731-1751	aldehydic	+	+	+	+	+	+
1750-1770	broth	+	+	+	+	+	+
1776-1781	fatty acid	nd	+	+	nd	+	+
1818-1822	mouldy	+	+	+	nd	nd	nd
1859-1873	fatty acid	+	+	+	+	+	+
1882-1896	medicinal	+	+	+	+	+	+
1906-1915	chemical	nd	+	+	+	+	+
1937-1943	floral	nd	+	nd	+	+	nd
1992-1998	fatty acid	+	+	+	nd	+	+
2033-2040	broth	+	+	+	+	+	+
2055-2082	caramel	+	+	+	+	+	+
2083-2124	caramel	+	+	+	+	+	+
2211-2232	plastic	+	+	+	+	+	+
2270-2277	woody	nd	nd	nd	+	+	+
2287-2296	sulfurous	nd	nd	nd	+	nd	+
2420-2425	floral	nd	+	+	nd	+	+
2508-2518	solvent	+	+	nd	+	+	nd

nd, not detected

The data are presented in Table VI. Altogether, the five fiber coatings enabled the detection of 45 odors: 38 with PDMS-DVB, 29 with CW-DVB, 35 with CAR-PDMS, 13 with PA and only three with PDMS. There were no unique odors extracted with PA and PDMS. Comparison of PDMS-DVB and CW-DVB revealed only one odor (sweet-apple sauce, RI 1828-1837) which was not detected when using PDMS-DVB. This odor, however, was detected when using the CAR-PDMS fiber, the latter extracting six additional odors (solvent, RI 816-830; fruity, RI 961-965; creamy, 988-1003; fruity, 1013-1014; green, RI 1086-1088; rotten, 1258-1259).

Table VI. Aroma Profiles of Extracted Cheddar Cheese Volatiles using Different Fiber Coatings (20°C, 12 h)(Table continued on next page)

RI	Descriptor	PDMS-DVB	CW-DVB	CAR-PDMS	PA	PDMS
736-745	sulfurous	+	nd	+	nd	nd
754-774	sulfurous	+	+	+	nd	nd
773-778	sulfurous	+	nd	+	nd	nd
781-790	sulfurous	+	+	+	nd	nd
816-830	solvent	nd	nd	+	nd	nd
913-929	solvent	+	+	+	nd	nd
961-965	fruity	nd	nd	+	nd	nd
988-1003	creamy	nd	nd	+	+	nd
1013-1014	fruity	nd	nd	+	nd	nd
1043-1048	fruity	+	+	+	nd	nd
1086-1088	green	nd	nd	+	nd	nd
1138-1142	fruity	+	nd	+	nd	nd
1240-1250	fruity	+	nd	+	nd	nd
1258-1259	rotten	nd	nd	+	nd	nd
1295-1304	floral	+	+	+	nd	nd
1309-1320	mushroom	+	+	+	nd	nd
1361-1373	fatty acid	+	nd	nd	nd	nd
1385-1391	sulfurous	+	+	+	+	nd
1411-1422	mouldy	+	+	+	nd	nd
1447-1458	fatty acid	+	+	+	nd	nd
1462-1466	aldehydic	+	nd	+	nd	nd
1469-1480	broth	+	+	+	+	nd
1518-1524	aldehydic	+	+	nd	nd	nd
1548-1558	aldehydic	+	+	+	+	nd
1600-1610	aldehydic	+	+	nd	nd	nd
1635-1662	fatty acid	+	+	+	+	nd
1662-1674	floral	+	+	nd	nd	nd
1686-1695	fatty acid	+	+	+	nd	nd

136

Table VI (Continued). Aroma Profiles of Extracted Cheddar Cheese Volatiles
using Different Fiber Coatings (20°C, 12 h)

RI	Descriptor	PDMS-DVB	CW-DVB	CAR-PDMS	PA	PDMS
1699-1713	broth	+	+	nd	+	nd
1720-1732	aldehydic	+	nd	nd	+	nd
1731-1751	aldehydic	+	+	+	nd	nd
1760-1767	broth	+	nd	+	+	+
1776-1781	fatty acid	+	nd	nd	nd	nd
1818-1822	mouldy	+	nd	nd	nd	nd
1828-1837	sweet	nd	+	+	+	nd
1859-1873	fatty acid	+	+	+	nd	nd
1882-1896	medicinal	+	+	+	nd	nd
1906-1915	chemical	+	+	nd	nd	nd
1937-1943	floral	+	+	+	nd	nd
1992-1998	fatty acid	+	+	+	nd	nd
2033-2040	broth	+	nd	nd	nd	nd
2055-2082	caramel	+	+	+	+	+
2083-2124	caramel	+	+	+	+	+
2211-2232	plastic	+	+	+	+	nd
2420-2425	floral	+	+	nd	nd	nd
2508-2518	solvent	+	+	+	+	nd

nd, not detected

Conclusion

The extraction yields of Cheddar cheese volatile compounds varied according to the fiber coating used. The different fiber coating also showed significant differences in selectivity towards the volatile compounds absorbed. There were distinct differences between aroma profiles of cheese volatiles extracted at 20°C for 12 h or at 60°C for 1 h. The importance of these differences to the contribution of Cheddar cheese aroma still needs to be assessed. The use of SPME-GCO, including dilution anlysis (30), may provide means to easily monitor specific cheese odors, especially during ripening.

Acknowledgments

The authors thank Mainland Cheese Factory (Dunedin, New Zealand) for supplying the Cheddar cheese samples. This research was supported in part by a grant from the New Zealand Dairy Research Institute.

Literature cited

1. Fox,P.F.; Singh,T.K.; McSweeney,P.L.H. *Chemistry of Structure-Function Relationship in Cheese*; Malin,E.L.;Tunick,M.H.,Eds; Plenun Press: New York, NY,1995,59-98.
2. Dacre, J.C. *J. Dairy Res.* **1955**, *22*, 219-223.
3. Walker, J.R. L.; Harvey, R.J. *J. Dairy Res.* **1959**,*26*,265-272.
4. Eaton,D.C. *Bio-process, production of flavor, fragance and colour ingredients*, Gabelman,A., Ed.; John Wiley and Sons Inc.: Ohio,1994,103-105.
5. Yang, W.T.; Min, D.B. *J. Food Sci.* **1994**,*59*,1309-1312.
6. Milo,C; Reineccius, G. *J. Agric. Food Chem.* **1997**,*45*,3590-3594
7. Aishima,T.; Nakai,S. *J. Food Sci.* **1987**,*52*,939-942.
8. Vandeweghe,P.; Reineccius,G. *J. Agric. Food. Chem.* **1990**,*38*,1549-1552.
9. Scarpellino,R.; Kosikowski,F.V. *J. Dairy Sci.* **1961**,*44*,10-15.
10. Libbey,L.M.; Bills,D.D.; Day,E.A. *J. Food Sci.* **1963**,*28*,329-334.
11. Day, E.A.; Libbey,L.M. *J. Food Sci.* **1964**,*29*,583-589.
12. Morris,H.A.; Angelini,P.; McAdoo,D.J.; Merrit,C.Jr. *J. Dairy Sci.* **1966**,*49*,710-716.
13. McGugan,W.A.; Howsam,S.G. *J. Dairy Sci.* **1962**,*45*,495-500.
14. Manning,D.J.; Robinson,H.M. *J. Dairy Res.* **1973**,*40*,63-75.
15. Manning,D.J.; Moore,C. *J. Dairy Res.* **1979**,*46*,539-545.
16. Price,J.C.; Manning,D.J. *J. Dairy Res.* **1983**,*50*,381-385.
17. Marsilli,R. *J. Dairy Sci.* **1985**,*68*,3155-3161
18. Dunn,H.C.; Lindsay,R.C. *J. Dairy Sci* **1985**,*68*,2853-2858.
19. Linn, J.C.; Jeon,I.J. *J. Food Sci.* **1985**,*50*,843-846.
20. Wong,N.P.; Park,O.W. *J. Dairy Sci.* **1968**,*51*,1768-1769.
21. Lamparski,D.; Klimes,I. *Flavor '81*; Schereier,R., Ed.; de Gruyter:Berlin,1981, 557-577.
22. Benkler,K.F.; Reineccius,G. *J. Food Sci.* **1980**,*45*,1081-1085
23. Liebich,H.M.; Douglas,D.R.; Bayer,E.; Zlatkis,A. *J. Chrom. Sci.* 1970,*8*,355-359
24. Gorecki,T.; Boyd-Boland,A.; Zhang,Z,; Pawliszyn,J. *Can. J. Chem.* **1996**,*74*,1297-1308.
25. Eisert,R.; Pawliszyn,J. *Critical Review in Analytical Chemistry*,**1997**,*27*,103-135
26. Zhang,Z.; Pawliszyn,J. *Anal.Chem.* **1993**,*66*,844-853
27. Chin,H.W.; Bernhard,R.A.; Rosenberg,M. *J.Food.Sci.***1996**,*61*,1118-1122&1128
28. Chin,H.W.; Rosenberg,M. *J.Food.Sci.***1997**,*62*,468-474
29. Clark,T.J.; Bunch,J.E. *J. Agric. Food Chem.* **1997**,*45*,844-849
30. Deiber, K.D.; Lavin, L.H.; Acree, T.E. *J. Agric. Food Chem.* **1999**,*47*,1616-1619

Chapter 12

Aroma Extract Dilution Analysis versus Aroma Extract Concentration Analysis

W. Grosch, R. Kerscher, J. Kubickova, and T. Jagella

Deutsche Forschungsanstalt für Lebensmittelchemie, Lichtenbergstrasse 4, D–85748 Garching, Germany (email: werner.grosch@Lrz.tu-muenchen .de

The potent odorants of boiled beef, Camembert cheese and black pepper were screened by AEDA and AECA. The latter procedure starts with gas chromatography/olfactometry (GCO) of the original extract which then is concentrated stepwise. After each step an aliquot is analyzed by GCO. This differs from AEDA in which the extract is concentrated to a small volume before GCO. It was found that the results of the two methods agreed aside from a few exceptions. Only the odor potencies of 4-hydroxy-2,5-dimethyl-3(2H)-furanone, 3-mercapto-2-pentanone and methional in boiled beef as well as α-phellandrene and limonene in black pepper were higher in AECA than in AEDA. Sensory studies of aroma models revealed that the results of AECA were correct.

Two analytical techniques, CHARM analysis (*1*) and AEDA (*2*), are applied for screening of potent odorants upon which identification experiments are focused. In both procedures, an extract obtained from the food is first concentrated to a smaller volume and then diluted stepwise with solvent, e.g. in a series of 1:1 dilutions. Then each dilution is analyzed by gas chromatography/olfactometry (GCO). As the dilution value obtained for each compound is proportional to its odor activity value (OAV, ratio of concentration to odor threshold) in air, the odorants can be arranged in order of their odor potency (*3*).

At the beginning of a CHARM analysis or an AEDA, the extract is strongly concentrated by distilling off the solvent. This procedure is performed to enrich the odorants for the identification experiments. However, it might lead to losses of odorants, for example by evaporation and by side-reactions, enhanced in the concentrated extract. Consequently, the odor potency of these compounds can be underestimated by GCO in comparison with the odor activity of those compounds, the levels of which are not reduced during concentration.

Aroma Extract Concentration Analysis (AECA) could be a milder alternative to AEDA (*4*).
This procedure starts with GCO of the original extract from which the non-volatile components have been removed. Then the extract is concentrated stepwise by distilling off the solvent, and after each step an aliquot is analyzed by GCO. To show the efficiency of the new method, it was compared with AEDA in studies on the aroma of boiled beef (*4*), French cheese (*5*) and black pepper (*6*).

Boiled Beef

Samples I and II of boiled beef were prepared for AEDA and AECA, respectively, as summarized in Table I and detailed in (*4*). At the beginning of AECA, an aliquot of sample II was subjected to GCO. As shown in Table II, only two odorants were detected: 2-furfurylthiol and furanone no. 2. In AEDA the odor activity of the thiol was only one dilution step lower than in AECA. This difference lies within the detection error of the method (*3*). However, the difference of 2 dilution steps which was found for furanone no. 2 might be too large to explain it by the detection error. Most likely a portion of furanone no. 2 was lost in AEDA. After concentration of sample II to 50 mL, 2-methyl-3-furanthiol, smelling like boiled beef, was detected in addition to compounds nos. 1 and 2. This result of AECA agreed with that of AEDA.

Table I. Preparation of the Samples for AEDA and AECA of Boiled Beef (*4*)

- Beef (forerib, 400 g) in water (300 mL) was boiled under pressure (8 x 10^4 Pa, 116°C) for 45 min.
- A portion (25 g) was frozen in liquid nitrogen, mixed with Na_2SO_4 and ground.
- Extraction with CH_2Cl_2 (2 x 100 mL, 4 h and 14 h).
- Distillation under high vacuum yielded a condensate of 180 mL.
- Adjustment of the condensate to 200 mL by addition of CH_2Cl_2 and division into halves (samples I and II, 100 mL each).
- Sample I was subjected to AEDA, which after concentration to 0.1 mL was performed.
- Sample II was subjected to AECA: an aliquot (1 μL) was analyzed by GCO. Sample II was concentrated stepwise. After each step an aliquot (1 μL) was analyzed by GCO.

A further concentration of sample II to 25 mL revealed no additional odorants. Only after reduction of the sample volume to 12.5 mL, three new odorants, nos. 4-6, were found. A comparison with the results of AEDA indicated that thiol no. 6 was lost in the latter procedure, as it was still detected after dilution to 3.1 mL.

After concentration of the aroma extract to 6.25 mL, a further 5 odorants, nos. 7-11, were perceived. The species-specific odorant 12-methyltridecanal (no. 10) (*7-9*), and a second furanone (sotolon, no. 11) appeared in this fraction. Methional (no.

Table II. Potent Odorants of Boiled Beef – Comparison AECA with AEDA (4)

No.	Odorant	Extract volume (mL)[a]	
		AECA	AEDA
1	2-Furfurylthiol	100	50
2	4-Hydroxy-2,5-dimethyl-3(2H)-furanone	100	25
3	2-Methyl-3-furanthiol	50	50
4	1-Octen-3-one	12.5	6.25
5	(E)-2-Nonenal	12.5	6.25
6	3-Mercapto-2-pentanone	12.5	3.1
7	Methional	6.25	1.6
8	Butyric acid	6.25	3.1
9	Guaiacol	6.25	3.1
10	12-Methyltridecanal	6.25	3.1
11	3-Hydroxy-4,5-dimethyl-2(5H)-furanone	6.25	3.1
12	Octanal	3.1	1.6
13	Nonanal	3.1	1.6
14	(E,E)-2,4-Decadienal	3.1	1.6
:			
22	Vanillin	3.1	3.1

[a] The extract volume at which the odorant was at first (AEDA) or at last (AEDA) perceived by GCO.

7) was the most labile of the five odorants, as its detection limit in AEDA at 1.6 mL was 4 times lower than in AECA.

The following experiments were performed to show whether the labile odorants furanone no. 2, 3-mercapto-2-pentanone and methional play a role in the aroma of boiled beef. The odorants screened by AECA and by gas chromatography/ olfactometry of headspace samples (GCOH) (8) were quantified in boiled beef and then their OAV were calculated on the basis of odor threshold values in water.

The results in Table III indicate that OAVs of 1000 and higher were found for furfurylthiol, 2-methyl-3-furanthiol, 12-methyltridecanal, nonanal and methanethiol. The latter belongs to the highly volatile odorants which were evaluated by GCOH. Also the OAV of furaneol was with 908 relatively high but those of mercapto-pentanone and methional were low.

Sixteen odorants in the concentrations occurring in boiled beef were dissolved in a base consisting of 10 % sunflower oil in a weak phosphate buffer of pH 5.7. Eleven assessors compared the odor profile of this aroma model with that of boiled beef.

The original sample smelled meaty, fatty, tallowy, sweet and caramel-like (Table IV). In addition, sulfurous, malty and roasty notes were to perceive. The fatty, tallowy note was intense because the sample contained with 8.8 percent a relative high amount of fat. The similarity score of 2.4 indicates that the odor profile of the aroma model was close to that of boiled beef (Table IV).

To investigate the contribution of each odorant occurring in the model to the odor of boiled beef, triangle tests were carried out. In each test, two samples containing the complete mixture of 16 odorants listed in Table V and one sample in which one odorant was lacking, were presented to eleven assessors. They had to find out the latter sample. The results in Table V reveal that nine odorants contribute at least significantly to the aroma of boiled beef. Furfurylthiol, furaneol, methyl-furanthiol but also mercaptopentanone and methional belong to this group. In a further triangle test the seven odorants whose contribution to the overall odor was not significantly established in experiments no. 10-16 (Table V) were omitted. When compared with the model containing the complete set of 16 odorants only 6 of the 11 assessors found the reduced model. This means that the aroma of the reduced

Table III. Concentrations and Odor Activity Values (OAV) of Potent Odorants of Boiled Beef

Odorant	Conc.[a] (µg/kg)	OAV[b]
2-Furfurylthiol	29	2900
4-Hydroxy-2,5-dimethyl-3(2H)-furanone (furaneol)	9075	908
2-Methyl-3-furanthiol	24	3429
1-Octen-3-one	9.4	188
(E)-2-Nonenal	32	128
3-Mercapto-2-pentanone	69	99
Methional	36	180
Butyric acid	7074	3
Guaiacol	4.3	2
12-Methyltridecanal	962	9620
3-Hydroxy-4,5-dimethyl-2(5H)-furanone	1.2	4
Octanal	382	546
Nonanal	1262	1262
Methanethiol (GCOH)[c]	311	1555
(E,E)-2,4-Decadienal	27	135
Dimethyl sulfide (GCOH)	105	350
(Z)-2-Nonenal	6.2	310
Acetaldehyde (GCOH)	1817	182
Methylpropanal (GCOH)	117	167

[a] The concentrations of the odorants were determined by stable isotope dilution assays (8, 10).

[b] OAVs were calculated by dividing the concentrations of the odorants by their odor threshold values in water (11).

[c] The high odor potency of the compound was established by gas chromatography/olfactometry of headspace samples (GCOH) (8).

Table IV. Odor Profile of Boiled Beef and Its Aroma Model[a]

Attribute	Beef	Model
	Intensity[b]	
Meaty	2.6 (0.4)	2.3 (0.5)
Fatty, tallowy	2.1 (0.5)	1.9 (0.3)
Sweet, caramel-like	1.7 (0.4)	1.8 (0.5)
Sulfurous	1.4 (0.6)	1.3 (0.6)
Malty	1.2 (0.6)	1.0 (0.7)
Roasty	1.1 (0.3)	1.0 (0.6)
Sweaty	0.4 (0.2)	0.3 (0.3)
Similarity[c]		2.4 (0.4)

[a]   Odor profile analysis was performed as reported for boiled chicken (*12*).
[b]   The intensity of the attributes was nasally scored on the scale 0 (absent) to 3 (strong). The results obtained by 11 panelists were averaged; the standard deviations are given in parentheses. No intensity of an attribute in the model was significantly different ($p < 0.05$) from the corresponding one in the odor profile of the beef sample (*17*).
[c]   Similarity score: 0 (no similarity) to 3 (identical with the original).

Table V. Odor of the Aroma Model for Boiled Beef as Affected by the Absence of One Odorant

Exp. no.	Odorant omitted[a]	Number[b]
1	2-Furfurylthiol	10***
2	4-Hydroxy-2,5-dimethyl-3(2H)-furanone	10***
3	3-Mercapto-2-pentanone	9***
4	Methanethiol	9***
5	Octanal	8**
6	2-Methyl-3-furanthiol	8**
7	Nonanal	8**
8	(E,E)-2,4-Decadienal	8**
9	Methional	7*
10	12-Methyltridecanal	6
11	Dimethyl sulfide	5
12	(Z)-2-Nonenal	4
13	Acetaldehyde	4
14	1-Octen-3-one	3
15	(E)-2-Nonenal	3
16	Methylpropanal	1

[a]   Odorant omitted in the aroma model.
[b]   Number of 11 assessors detecting the reduced model.
  *   Significant ($p < 0.05$),
  **   highly significant ($p < 0.01$).
  ***   very highly significant ($p < 0.001$).

model containing nine odorants does not differ significantly from that of the complete model. This allows the conclusion that these volatiles are the character impact flavor compounds of boiled beef.

The aroma model in which mercaptopentanone was lacking (no. 3 in Table V) was correctly identified by 9 of the 11 assessors. This result confirmed that of AECA indicating that mercaptopentanone is an important contributor to boiled beef aroma. In AEDA, however, the odor potency of this thiol was underestimated. Also in the cases of furaneol and methional the higher odor potency found by AECA in comparison to AEDA was in agreement with the results of the omission tests listed in Table V.

The higher odor potency that was found for 1-octen-3-one and (E)-2-nonenal in AECA and AEDA (Table II) was not confirmed by the calculation of OAV (Table III). The latter result agreed with the sensory experiments in which the two carbonyl compounds were not missed (Table V). However, also 12-methyltridecanal did not significantly affect the aroma of the model although its OAV was very high. Obviously, it did not successfully compete with octanal, nonanal and decadienal for the stimulation of the fatty note in the odor profile of boiled beef. However, as shown in a previous study (8), 12-methyltridecanal belongs to the key odorants of the juice that was released during stewing of lean beef. As it originates from membranes of the muscle tissue and not from the depot fat, its concentration is also high in lean beef.

French Cheese (Camembert)

The aroma of the French cheese Camembert was also analyzed with the two screening methods for potent odorants (5). Camembert is a soft cheese with a mushroom-like odor note, which according to Dumont et al. (13) is caused by 1-octen-3-ol.

For comparison of the results of the two screening experiments the ranking in the odor activity of the volatiles is presented in Table VI. 3-Methylbutanal (no. 1) and methional (no. 2) showed the highest odor activity in AECA but also in AEDA. After reduction of the volatile fraction of the cheese sample to the half by distilling off the solvent, 5 further odorants were perceived by GCO. The mushroom-like smelling odorants 1-octen-3-ol (no. 4a) and 1-octen-3-one (no. 4b) belong to this group. They were not separated on the unpolar capillary SE-54, which was used here. Further concentration of the cheese extract provided phenylethyl acetate (no. 8) and isovaleric acid (no. 9) as important odorants. In contrast to AECA the acid fraction was separately analyzed in AEDA.

A comparison of the rankings in the odor odor potency of the volatiles reveals that the difference between the concentration method and the dilution method is only one unit and this is within the limit of error.

Using unripened cheese material as base an aroma model was prepared for Camembert (14). The 11 odorants, which are labelled by an asterisk in Table VI,

Table VI. French Cheese (Camembert) – Ranking of Potent Odorants in AECA and AEDA

No.	Odorant	Ranking[a]	
		AECA	AEDA
1	3-Methylbutanal*	1	1
2	Methional*	1	1
3	2,3-Butanedione*	2	3
4a/b	1-Octen-3-one*/1-octen-3-ol*	2	3
5	2-Undecanone	2	2
6	δ-Decalactone*	2	3
7	Butyric acid*	2	-[b]
8	Phenylethyl acetate*	3	4
9	Isovaleric acid*	3	-[b]
10	2-Acetyl-1-pyrroline*	4	4
11	Dimethyl trisulfide*	4	5
12	Phenylacetaldehyde	4	4
13	Unknown (RI 1266 on SE-54)	4	5
14	Unknown (RI 1081 on SE-54)	5	5
15	p-Cresol	5	>5
16	(Z)-6-Dodecen-γ-lactone	5	>5
17	γ-Dodecalactone	5	n.d.

[a] Ranking of odor activity: 1, high ... 5, low.
[b] The acids were separately analyzed.
n.d., not detectable.
* Component of the aroma model.

were added to the base in the concentrations found in the cheese. Undecanone (no. 5) was not used because its OAV, which was calculated on the basis of the odor threshold in oil, was lower than one (*15*).

Besides the 11 odorants labelled in Table VI methanethiol, dimethylsulfide, acetaldehyde and methylene bis(methylsulfide), whose high odor potency was established by GCOH, were added to the aroma model. Also the taste compounds, which were evaluated in a separate study (*14*), were included.

The odor profiles of the base and of the Camembert were compared with that of the aroma model (Table VII). The base, an unripened cheese, smelled intensely buttery and slightly malty and sweet. Addition of the mixture of flavor compounds to the base shifted the nasally and retronasally evaluated odor profile in the direction of Camembert cheese. The intensity of the buttery note was reduced. This note and the sulfurous, garlic-like note agreed with the corresponding odor impressions of the Camembert cheese sample. The mushroom-like aroma was perceived in the aroma model, but its intensity was somewhat lower, when the odor profile was nasally assessed.

In summary, the odor profile of the aroma model was close to that of genuine Camembert. This indicates that AECA and AEDA have shown the odorants causing the aroma of this type of cheese. However, omission experiments analogous to those discussed for boiled beef have to indicate whether all of the odorants present in the model contribute to the aroma of Camembert.

Table VII. Odor Profile of the Base, the Aroma Model and a Camembert Cheese Sample (Cam)[a]

| Attribute | Intensity[b] | | | | | |
| | Base | | Model | | Cam | |
	n	rn	n	rn	n	rn
Buttery	2.1	1.6	0.9	1.0	1.1	1.2
Sulfurous, garlic-like	0	0	2.5	2.1	2.2	2.0
Mushroom-like	0	0	1.0	1.5	1.6	1.7
Floral	0	0	0.3	0	0.4	0
Malty, sweet	0.6	0.5	0.6	0.4	0.4	0.4
Pungent, sweaty	0.3	0.3	1.2	1.0	1.0	0.7
Ammoniacal	0	0	0.6	0.5	0.8	0.6

[a]   The odor profile was nasally (**n**) and retronasally (**rn**) assessed.
[b]   Intensity rating scale: 0 (absent) to 3 (strong). The results obtained by seven panelists were averaged.

Black Pepper

In this case the difference between the results of AECA and AEDA was greater than in the example discussed before. As shown in Table VIII, the greatest differences were found for α-phellandrene (no. 3) and limonene (no. 5) whose odor activities were much higher in AECA than in AEDA. The odor activity of cineole

Table VIII. Black Pepper (*Piper nigrum*) – Ranking of Potent Odorants in AECA and AEDA (6)

| No. | Odorant | Ranking[a] | |
		AECA	AEDA
1	Linalool	1	1
2	α-Pinene	1	3
3	α-Phellandrene	1	>6
4	Myrcene	2	3
5	Limonene	2	6
6	1,8-Cineole	?	?
7	Unknown	4	?
8	Sabinene/ß-pinene	5	>6
9	(E,Z)-1,3,5-Undecatriene	5	>6
10	Wine lactone	5	>6
11	Ethyl cinnamate	5	>6
12	ß-Ionone	5	>6
13	Germacrene B	5	>6

[a]   Ranking of odor activity: 1, high ..... 6, very low.

(no. 6) could not be evaluated because it was not clearly separated by gas chromatography from the high amount of limonene.

An aroma model using starch as base was developed for black pepper on the basis of quantitative data (16). Then the composition of the obtained aroma model was varied to gain an insight into the character impact flavor compounds.

A cut of the results showing the contribution of the monoterpenes is summarized in Table IX. In exp. no. 1 α-phellandrene was clearly missed by the assessors. In this case but also in that of limonene, which is shown in the following experiment, the result of AECA was correct. Both monoterpenes are key odorants of black pepper.

Table IX. Odor of the Aroma Model for Black Pepper as Affected by the Absence of Various Monoterpenes

Exp. no.	Odorant(s)[a]	Number[b]
1	(S)-α-Phellandrene	10[c]
2	Limonene	8[c]
3	(-)-α-Pinene, (+)-α-pinene, (-)-ß-pinene, myrcene, 1,8-cineole	9[c]
4	(-)-α-Pinene, (+)-α-pinene, (-)-ß-pinene	7[c]
5	Myrcene	8[c]
6	1,8-Cineole	6
7	Linalool	7[c]

[a]   Two aroma models containing the complete set of 15 odorants and one reduced model were presented in each triangle test (16).
[b]   Number of 10 assessors detecting the reduced model.
[c]   Significant result (p < 0.05).

Conclusions

Most of the potent odorants agree when the volatile fraction of a food is analyzed by AEDA and AECA.

Of the odorants evaluated in boiled beef, Camembert cheese and black pepper, only 4-hydroxy-2,5-dimethyl-3(2H)-furanone, 3-mercapto-2-pentanone and methional in boiled beef as well as α-phellandrene and limonene in black pepper were exceptions. Their high odor potencies in the corresponding volatile fractions were correctly estimated by AECA but not by AEDA.

References

1.   Acree, T.E.; Bernard, J.; Cunningham, D.G. *Food Chem.* **1984**, *14*, 273-286
2.   Ullrich, F.; Grosch, W. *Z. Lebensm. Unters. Forsch.* **1987**, *184*, 277-282

3.   Grosch, W. *Trends Food Sci. Technol.* **1993**, *4*, 68-73
4.   Kerscher, R.; Grosch, W. *Z. Lebensm. Unters. Forsch.* **1997**, *204*, 3-6
5.   Kubickova, J.; Grosch, W. *Int. Dairy Journal* **1997**, *7*, 65-70
6.   Jagella, T.; Grosch, W. *Euro Food Res. Technol.* **1999**, *209*, 16-21
7.   Guth, H.; Grosch, W. *Lebensm. Wiss. Technol.* **1993**, *26*, 171-177
8.   Guth, H.; Grosch, W. *J. Agric. Food Chem.* **1994**, *42*, 2862-2866
9.   Guth, H.; Grosch, W. *Z. Lebensm. Unters. Forsch.* **1995**, *201*, 25-26
10. Kerscher, R.; Grosch, W. *J. Agric. Food Chem.* **1998**, *46*, 1954-1958
11. Kerscher, R.; Grosch, W. *Proceedings of the 9th Weurman Flavour Research Symposium*, Technical University Munich, Freising, 22-25 June 1999 (in press)
12. Kerler, J.; Grosch, W. *Z. Lebensm. Unters. Forsch.* **1997**, *205*, 232-238
13. Dumont, J.P.; Rogers, S.; Cerf, P.; Adda, J. *Lait* **1974**, *54*, 501-516
14. Kubickova, J.; Grosch, W. *Int. Dairy J.* **1998**, *8*, 11-16
15. Kubickova, J.; Grosch, W. *Int. Dairy J.* **1998**, *8*, 17-23
16. Jagella, T.; Grosch, W. *Eur. Food Res. Technol.* **1999**, *209*, 22-26
17. Kaiser, R.; Gottschalk, G. *Elementare Tests zur Beurteilung von Messdaten*; Bibliographisches Institut: Wissenschaftsverlag, Mannheim, Germany, **1972**, pp. 18-27

Chapter 13

Selecting Standards for Gas Chromatography–Olfactometry

Jane E. Friedrich, Terry E. Acree, and Edward H. Lavin

New York State Agricultural Experiment Station, Department of Food Science and Technology, Cornell University, Geneva, NY 14456

Gas chromatography - olfactometry (GC/O) is commonly used to identify odor-active chemicals in a sample extract or head-space. A persistent problem with GC/O data is the irreproducibility associated with human detection of an odorant region. To account for this irreproducibility researcher's often average data from two or more sniffers, use standard mixtures of 6 to 10 odorants to screen subjects, or both. However, the current odorant mixtures do not cover all the known anosmias or the odor classes humans experience. This paper describes an odorant mixture that has been formulated to stimulate all human olfactory receptor proteins (OP's) on the theory that OP's are specific for aroma categories. Compounds were chosen based on the following criteria: those known to be associated with specific anosmia, that gave baseline separation (RI OV101), that were stable (stability in the mixture and chromatographically), commercially available, and low in toxicity. The standard odorant mixture was formulated and analyzed by a reference individual to create a benchmark for further GC/O studies. Using the benchmark values for the aroma categories we are able to characterize each sniffer's olfactory acuity and quantitatively measure the olfactory acuity of each sniffer relative to the reference individual.

Background

Gas chromatography - olfactometry (GC/O) has been applied to problems in flavor analysis for the past 25 years. During this time the large variability associated with the human olfactory system has been well documented. This shows a large difference in the ability of humans to detect odors independent of cognition [1-5]. The difference in human responsiveness to odors is that at the transduction level the pattern of odor detectors, i.e. olfactory receptor proteins (OP's) in the olfactory system, is not the same in all humans.

In the late 1960's Amoore observed that there were individual differences in odor thresholds. Individual odor thresholds that were two standard deviations or greater than the mean response of a population were termed specific anosmic [6]. Those responses that were two standard deviations or less than the mean response of a population were termed hyperosmic. The phenomenon of specific anosmia is easily observed by GC/O and biases the results of GC/O measurements [7].

Standard mixtures of 6 to 10 odorants have been used to screen subjects for anosmia; however, they do not cover all known anosmias or the odor classes humans experience. Averaging the responses of two to six subjects in GC/O analysis has also been used [7] to minimize bias caused by anosmia. Developing one mixture of odorants representing all specific anosmias would enhance this approach and justify the use of a smaller subject sample.

Variation of Human Olfactory Acuity

One reason for the large variation associated with olfactory acuity is due to the phenomenon of specific anosmia. Specific anosmia is a 'smell blindness' or insensitivity to the odor of a chemical or group of chemicals in people with otherwise normal olfactory sensitivity. Specific anosmia poses a challenge to researchers when it biases the data produced in sensory experiments. Furthermore, if it is true that the perception of mixtures, the normal state of human olfaction, is a function of component ligand binding to OP's then dysosmia, the distortion of olfactory perception, is necessarily a consequence of specific anosmia. It is possible that most of the variation in human olfaction is a result of specific anosmia and its consequent distortion of the perception of odorant mixtures. Specific anosmia is quantitatively defined as responses less than two standard deviations from the population mean or the most sensitive group of a bimodal distribution [6]. Eighty-nine specific anosmias have been reported affecting between 3 and 47% of the population [8-11]. Of all the specific anosmia's reported the most thoroughly studied is that to androstenone, for which a genetic basis has been well-established [12].

Deviations in ability to smell include hyposmia, a decrease in sensitivity; anosmia, the lack of sensitivity; hyperosmia, a heightened sensitivity; and cacosmia and parosmia, abnormal qualitative perceptions of odors [2, 13]. There are two types of anosmia: specific and general. Specific anosmia is defined as an individual odor threshold that is two standard deviations or greater than the mean response of a population thus it is associated with quantitative changes in odor assessment. Some of these olfactory dysfunctions have been used as indictors of disease [14, 15]. Adaptation studies have also associated specific anosmia with qualitative changes in odor assessment [16, 17]. General anosmia, the loss of all odorant detection, is usually caused by head injuries but genetics has also been shown to be a contributor [13]. For example, the heritable loss of olfactory function occurring in patients with X-linked Kallman's syndrome due to mutation in an N-CAM-like protein apparently essential for neuronal migration and olfactory bulb formation causes a general anosmia [18]. Changes in the sense of smell during the life of an individual have also been associated with certain pathologies [19, 20]. Although olfactory sensitivity has been found to change with age there is little evidence that specific anosmia is associated with aging.

Aside from specific anosmia, research on smell shows strong differences between individuals. In 1988 Marin et al. [1] used GC/O to study the variation in odor detection thresholds to seven compounds analyzed by four groups of individuals cross-classified by age and sex. Subjects were selected for their "normal" olfactory acuity. It was found that the olfactory diversity measured by GC/O couldn't be attributed to age and sex differences alone. None of the thresholds were influenced by sex and only one, *l*-carvone, was influenced by age. This study was then repeated with another group of individuals producing the same result. A study by Lawless et al. in 1994 showed that the mean individual odor threshold of 53 individuals for diacetyl varied by over a factor of 256. Wide individual differences in odor thresholds were also shown for *l*-carvone and cineole [21, 22].

Detection of specific anosmia in humans

Currently, there is no quick precise method of quantitatively measuring the olfactory acuity of humans using sensory protocols. Specific anosmias are formally established by threshold testing but have also been discovered by researchers engaged in unrelated studies. For example, isobutryic acid, which has the "terrible odor of dirty socks or unclean goat" to most individuals, had a "very pleasant fruity odor – like apples" to two individuals at the Western Regional Research Laboratory, Albany, CA in 1967. Preliminary experiments revealed that these individuals had a specific anosmia to isobutyric acid and they were detecting the by-products and impurities usually found in commercial samples of isobutyric acid [6]. This work led to a more formal methodology to screen for specific anosmia using two target odors embedded in three blanks at each ascending concentration, first published in 1979 [9]. Threshold methods are time consuming, produce uncontrolled flow rates, and random inspiration patterns and require diligence to determine the range of doses needed to bracket the thresholds of different odorants. They are further limited by the number of odorants that can be tested in one session being less than 10. Adaptation, the loss of sensitivity with exposure to potent concentrations of an odorant, is also a common occurrence with threshold testing. Other similar tests have been developed including the UCONN test [3, 5, 23], the odor identification test [3-5], and the UPSIT test (40 scratch and sniff odors) [23], however, none of these tests can produce precise thresholds of 40 different odorants in one or two days.

Delivering a large number of chemicals (~40) to subjects for threshold testing can be easily achieved with a gas chromatograph coupled to a dynamic olfactometer or a GC/O. Over the last twenty years the design of the GC/O has evolved to preserve laminar flow at every stage from the chromatograph to contact with the sniffer, resulting in resolution at the sniff port approaching that of the capillary column itself [24]. Instead of a square pulse of odorant as in threshold testing methods, GC/O delivers the stimulant in a gaussian gradient that is very reproducible and only 2 or 3 seconds in duration thus presenting subjects with a very pure dose of odorant. By repeatedly sniffing the sample over a series of dilutions an estimate of the threshold for each compound can be determined. GC/O is quick (each run takes 30 minutes or less) and it is possible to test 20 to 30 odorants at one time. Therefore in two days a subject could be completely screened for any specific anosmia. It is an ideal method for testing a large number of odorants in a short time as long as the compounds are stable under gas chromatographic conditions, separable from each other and adjusted to equal potency. This can then be used to test for anosmias and hyperosmias based on the deviation from the standard potencies.

Objective

The objective of this research is to design a standard set of chemicals to screen for specific anosmia in humans using gas chromatography – olfactometry (GC/O). A comprehensive standard set of odorants was formulated to have equal odor potency and stimulate all OP's on the basis that OP's are specific for an aroma category. To stimulate all OP's the standard odorant mixture includes odorants from each aroma genera. Using GC/O to deliver the standard set provides a method that is quick, reproducible, and precise. The standard set of odorants provides an effective tool for screening for specific anosmia and selection of subjects for GC/O testing. The results will give us the basic knowledge we need to understand the effects specific anosmia has on standard GC/O analysis and provide a method to standardize GC/O-derived potency data collected from natural products, foods and environmental smells.

Experimental

The standard odorant mixture was developed so that all potential olfactory receptor proteins are stimulated, therefore the primary consideration for odorant selection was aroma class. An aroma genus of 26 aroma genera has been created based on ASTM DS66, a list of 23 food aroma categories [25] plus 3 non-food aroma categories. Using this aroma genus compounds can be classified into aroma categories. The 26 genera are; maillard, dairy, edible oil, fermented, fishy, shellfish, berry, citrus, pome fruit, stone fruit, tropical, grain, cured meat, fabricated meat, processed meat, raw meat, herbs, peppers, roots spices, seed spices, sweet spices, aromatic, vegetable, floral, animal, and mineral [26].

Formulation of the Standard Solution Set

A standard set of 40 odorants, approximately two odorants for each genera, should be sufficient to cover all aroma categories. Further criteria for compound selection include; compounds with known specific anosmia, compounds that will give baseline separation (RI OV101), compound stability, compound availability, and non-toxicity. Chemicals were obtained from commercial sources (Sigma, Aldrich, etc.) and made into solution in either Freon 113™ or ethyl acetate to obtain non-polar and polar solutions, respectively. The polar, ethyl acetate, standard set is shown below in Table I. Two solutions are used to account for solubility and good chromatographic separation of the chemical standards. This paper reports the analysis and results of the polar standard solution.

Screening

Standard solutions were analyzed using a GC/O system (CharmAnalysis™) on a modified HP 6890 GC (Datu, Inc., Geneva, NY) and sniffed using a HP-1 (15 m x 0.32 mm) column. The oven temperature was programmed from 35 to 250 °C at 6 °C/min. A dilution series of the standard set was performed and all solutions were sniffed in duplicate. The retention time of each odorant was converted to Kovats indices using 7-28 carbon normal paraffins. A reference individual analyzed all solution sets to determine appropriateness (odor quality, chromatographic separation,

Table I. Polar standard set of odorants.

Retention Index (OV101)	Compound	Genera	CAS #
0844	2-methyl-3-furanthiol	meat	28588-74-1
0920	α–pinene	herbs	80-56-8
0995	methyl cyclopentenolone	maillard	765-70-8
1005	hexanoic acid	dairy	142-62-1
1034	2,5-dimethyl-4-hydroxy-3(2H)-furanone	berry	3658-77-3
1048	m-cresol	mineral	108-39-4
1070	Maltol	maillard	118-71-8
1216	γ-octalactone	tropical	104-50-7
1304	hydrocinnamic acid	fermented	501-52-0
1346	Vanillin	aromatic	121-33-5
1445	ethyl 2,4-decadienoate	pome	3025-30-7
1498	raspberry ketone	berry	5471-51-2
1673	7-methoxycoumarin	stone	531-59-9
1751	Nootkatone	citrus	4674-50-4

etc.). The concept of the reference individual is to create a benchmark value for all aroma classes. By creating benchmark values for all aroma classes we are able to quantitatively measure the acuity of individuals and quantitatively compare results from different laboratories.

A preliminary polar odorant mixture was analyzed by the reference individual. Prior screening using the procedure of Marin et al. [1] indicated that the reference individual has no known specific anosmia. All dilutions were sniffed in duplicate and the retention index of each odorant was converted to Kovats indices using 7-18 carbon normal paraffins. The concentration of each odorant was then adjusted to equal odor potency creating benchmark values for each odorant.

Analysis

Three subjects analyzed the polar standard set of odorants. The olfactory acuity of all subjects was screened prior to analysis according to the methods of Marin et al. [1] and GC/O analysis was performed as described above in the subsection titled screening. Charm values were converted to modified odor spectrum values (OSV_m) using a Steven's law exponent of 0.5 for olfaction. Odor spectrum values (OSV) are indicators of odor potency; they are transformed normalized Charm values and are independent of concentration. A modified odor spectrum value (OSV_m) excludes the anomalous responses, the Charm values of hyperosmic and anosmic responses. Therefore, an OSV_m is a transformed normalized Charm value that is also normalized for each individual. An OSV_m chromatogram is made by plotting response (OSV_m) versus retention index (RI) and is representative of the pattern of odorants at the olfactory epithelium [24] for each individual.

Results and Discussion

Olfactory acuity was determined using a variation of John Amoore's definition of specific anosmia [6]. If a response deviates more than two standard deviations from the mean natural log Charm response of the reference individual it is considered a specific anosmic response. And if the response deviates less than two standard deviations from the mean natural log Charm response of the reference individual it is considered a hyperosmic response. The mean natural log Charm is used because it has been shown that performing a log transformation of Charm data stabilizes the dispersion of the data and allows statistical tests to be applied with more justification [28].

The modified odor spectrum value (OSV_m) chromatograms of the polar standard set as analyzed by the reference individual and test subjects is shown in Figure 1. The responses of the reference individual are of equal odor potency indicating that the standard set is a good approximation for the reference individual. The pattern of responses to the polar standard set of odorants differed for all subjects tested, showing the large variability of human olfaction.

Assignments of olfactory acuity for each odorant were made using the benchmark values created by the reference individual. Peaks **a** through **f** of Figure 1 indicate odorants for which one or more of the test subjects responses were two standard deviations from the mean response of the reference individual. For example, subject 1 displayed a specific anosmia to α-pinene (**b**), being unable to detect α–pinene at levels up to 100 ng, and was hyperosmic to 2,5-dimethyl-4-hydroxy-3(2H)-furanone (**e**), detecting this compound at levels as low as 50 pg corresponding to a modified odor spectrum value of 381. Both responses were two standard deviations from the mean ln(Charm) of the reference individual. The responses of subject 2 did not deviate more than two standard deviations from those of the reference individual therefore

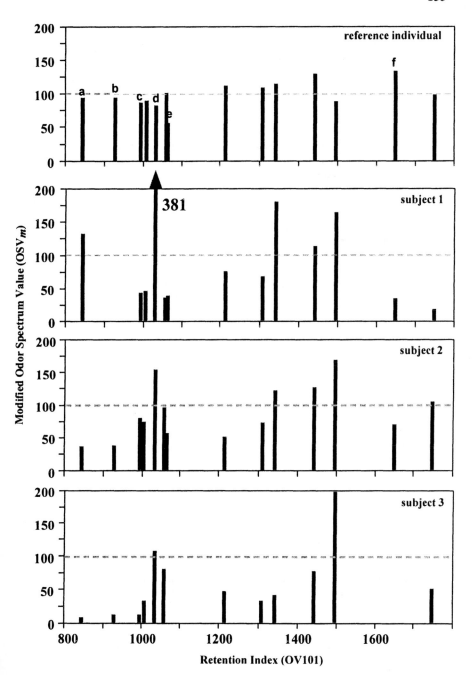

Figure 1. OSV_m chromatograms of the polar standard solution set (RI versus OSV_m).

they are not specific anosmic or hyperosmic to any of the odorants in the polar standard set of odorants. Subject 3 was found to display specific anosmia's to 2-methyl-3-furanthiol (**a**) at 23 pg, α-pinene (**b**) at 137 ng, methyl cyclopentenolone (**c**) at 865 pg, maltol (**d**) at 4 ng, and 7-methoxycoumarin (**f**) at 3 ng. Displaying specific anosmias to methyl cyclopentenolone and maltol, both of which are representatives of the maillard aroma genus, supports the theory that OP's are specific for an aroma category.

This chemical standard set provides the screening tool needed for GC/O aroma research and standardization of GC/O allowing the comparison of data produced in different laboratories by different subjects. At present, there is not a universal method of screening individuals prior to GC/O analysis. Many researchers currently screen their subjects with standard mixtures of 6 to 10 odorants. However, the current standards do not cover all the known anosmias or all the odor classes humans experience. Therefore, the development of a comprehensive standard set of odorants formulated to have similar odor potency and (presumably) stimulate all olfactory receptor proteins (OP's) provides an effective tool to screen subjects for GC/O analyses. This will not only provide a rapid screening method for specific anosmia in humans but also quantitatively determine the olfactory acuity of individuals. Deciding how to apply this tool requires data from a larger sample of subjects. Furthermore using a standard set of odorants that is appropriately designed to stimulate all potential olfactory receptor proteins could provide insight into the mechanism of human olfaction.

Literature Cited

1. Marin, A.B., T.E. Acree and J. Barnard, *Variation in odor detection thresholds determined by charm analysis.* Chemical Senses, 1988. **13**(3): p. 435-444.
2. Smith, D.V. and H.J. Duncan, *Primary Olfactory Disorder: Anosmia, Hyposmia, and Dysosmia,* in *Science of Olfaction,* M.J. Serby and K.L. Chobor, Editor. 1992, Springer-Verlag: New York. p. 439-466.
3. Cain, W.S., J. Gent, F.A. Catalanotto and R.B. Goodspeed, *Clinical evaluation of olfaction.* Am. J. Otolaryngol., 1983. **4**: p. 252-256.
4. Cain, W.S. and J.F. Gent, *Use of odor identification in clinical testing of olfaction,* in *Clinical Measurement of Taste and Smell,* H.L. Meiselman and R.S. Rivlin, Editor. 1986, Macmillan: New York. p. 170-186.
5. Cain, W.S., J.F. Gent, R.B. Goodspeed and G. Leonard, *Evaluation of olfactory dysfunction in the Connecticut Chemosensory Clinical Research Center.* Laryngoscope, 1988. **98**: p. 83-88.
6. Amoore, J.E., D. Venstrom and A.R. Davis, *Measurement of specific anosmia.* Perceptual and Motor Skills, 1968. **26**: p. 143-164.
7. Abbott, N., P. Etievant, S. Issanchou and L. Dominique, *Critical evaluation of two commonly used techniques for the treatment of data from extraction dilution sniffing analysis.* J. Agric. Food Chem., 1993. **41**(10): p. 1698-1703.
8. Amoore, J.E., *Specific anosmia and the concept of primary odors.* Chemical Senses and Flavor, 1977. **2**: p. 267-281.
9. Amoore, J.E., *Directions for preparing aqueous solutions of primary odorants to diagnose eight types of specific anosmia.* Chemical Senses and Flavour, 1979. **4**(2): p. 153-161.
10. Amoore, J.E., *Specific Anosmias,* in *Smell and Taste in Health and Disease.* 1991, p. 655-664.
11. Pelosi, P. and R. Viti, *Specific anosmia to l-carvone: the minty primary odor.* Chem. Senses Flavor, 1978. **3**(3): p. 331-337.

12. Wysocki, C.J. and G.K. Beauchamp, *Ability to smell androstenone is genetically determined.* Proc. Nat. Acad. Sci. USA, 1984. **81**: p. 4899-4902.
13. Engen, T., *The Perception of Odors.* 1982, New York: Academic Press.
14. Gregson, R.A.M., M.L. Free and M.W. Abbott, *Olfaction in korsakoffs, alcoholics and normals.* Br. J. Clin. Psychol., 1981. **20**: p. 3-10.
15. Gregson, R.A.M. and D.E.R. Smith, *The clinical assessment of olfaction: differential diagnosis including Kallman's syndrome.* J. Psychosom. Res., 1981. **25**(3): p. 165-174.
16. O'Connell, R.J., D.A. Stevens and L.M. Zogby, *Individual differences in the perceived intensity and quality of specific odors following self- and cross-adaptation.* Chemical Senses, 1994. **19**(3): p. 197-208.
17. O'Connell, R.J.O., D.A. Stevens, R.P. Akers, D.A. Coppola and A.J. Grant, *Individual differences in the quantitative and qualitative responses of human subjects to various odors.* Chem. Senses, 1989. **14**(2): p. 293-302.
18. Reed, R.R., *The molecular genetics of olfaction,* in *Olfaction and Taste. A Century for the Senses.,* G. Salvadori, Editor. 1997, Allured Publishing Corporation: Carol Stream, IL. p. 137-148.
19. Halpern, B.P., *Environmental factors affecting chemoreceptors: an overview,* in *Toxicology of the eye, ear, and other special senses,* A.W. Hayes, Editor. 1985, Raven Press: New York.
20. Schiffman, S., *Taste and smell in disease (Part 1).* New Engl. J. Med., 1983. **38**(21): p. 1275-1279.
21. Lawless, H.T., M.J Antinone, R.A. Ledford and M. Johnston, *Olfactory responsiveness to diacetyl.* Journal of sensory studies, 1994. **9**: p. 47-56.
22. Lawless, H.T., C.J. Corrigan Thomas and M. Johnston, *Variation in odor thresholds for l-carvone and cineole and correlations with suprathreshold intensity ratings.* Chem. Senses, 1995. **20**: p. 9-17.
23. Doty, R.L., P. Shaman, S.L. Applebaum, R. Giberson, L. Siksorski and L. Rosenberg, *Smell identification ability: changes with age.* Science, 1984. **226**: p. 1441-1443.
24. Acree, T.E., *GC/Olfactometry, GC with a sense of smell.* Analytical Chemistry, 1997. (69): p. 170A-175A.
25. Civille, G.V. and B.G. Lyon, *Aroma and flavor lexicon for sensory evaluation: terms, definitions, references, and examples.* 1996, West Conshohocken: American Society of Testing and Materials.
26. Acree, T.E. and H. Arn, *Flavornet. Gas chromatography – olfactometry (GCO) of natural products.* 1997,
27. Acree, T.E. and J.B. Barnard, *Gas chromatography-olfactometry and CharmAnalysis^{TM},* in *Trends in Flavour Research,* H. Maarse and D.G. van der Heij, Editor. 1994, Elsevier Science B.V.: Amsterdam. p. 211-220.

Chapter 14

Limitations in the Use of Odor Activity Values to Determine Important Odorants in Foods

Valérie Audouin, Florence Bonnet, Zata M. Vickers, and Gary A. Reineccius

Department of Food Science and Nutrition, University of Minnesota, 1334 Eckles Avenue, Saint Paul, MN 55108

The aim of this study was to investigate if OAVs are appropriate tools for predicting the perceived intensity of odorants and their contribution to the overall aroma. Two similar studies with two sensory panels were conducted. The detection thresholds and the relation of perceived intensity to concentration of several aroma compounds were determined. The threshold values of these compounds were different but varied considerably among and within individuals. The steepness of the slopes and shapes of the odor intensity functions were different for the odorants. A consequence of this was that the OAVs (ratio of the concentration of an odorant to its threshold concentration) for the panels were not useful measures of intensities. In a final experiment, we had subjects rate the contribution of two compounds to the overall intensity of a binary mixture. The OAVs were not good indicators of the percent contribution to the overall intensity of a mixture

The aroma of a food product is produced by simultaneous perception of many volatile compounds. However, only a limited number of them have any significance in flavor determination. Therefore, some criteria are needed to distinguish the compounds that contribute significantly to the overall aroma.

Patton and Josephson (*1*) first proposed estimating the importance of a flavor compound to a food based on the ratio of the compound concentration in the food to its threshold concentration in that food. This ratio is known now as the odor activity value (OAV) (also as: odor value, odor unit, flavor unit, and aroma value). This ratio indicates by how much the actual concentration of a compound exceeds its sensory threshold. They suggested that "compounds which exceed the threshold level in a food are significant in its flavor, whereas those occurring below threshold are not". Patton and Josephson (*1*) proposed this method as a guidance tool "that may not hold in some instances". The OAV concept was applied to mixtures by Guadagni *et al.* (*2*) who suggested that if the perceived intensity of odorants in a mixture is additive, the relationship between OAV of single components in the mixture and the OAV of that mixture was:

$$OAV_{i1} + OAV_{i2} + ... + OAV_{in} = OAV_m$$

where i1 ... in : compound 1, ..., n in mixture m.

   Thus, the relative contribution of a compound to a mixture could be described as the ratio of its OAV to the OAV of the mixture. Guadagni *et al.* (*2*) noted that this implied nothing about the odor quality of the final mixture and nothing about the relationship between the stimulus concentration and sensation above threshold.
   Since the introduction of the OAV concept, gas-chromatography/olfactometry (GCO) and OAV have been extensively used to screen for significant odorants in food. Two major screening procedures for determining the key odorants in food are based on this concept: the Aroma Extract Dilution Analysis (AEDA) developed by Ullrich and by Grosch (*3*) and a recent variation, the Aroma Extract Concentration Analysis (AECA) by Kerscher and Grosch (*4*) and Charm Analysis developed by Acree and coworkers (*5*). These two methods evaluate by GCO a dilution (or concentration) series of the original aroma extract from a particular food and attempt to rank the key odorants in order of potency. The highest dilution at which a substance is smelled is defined as its dilution value. The dilution value is proportional to the OAV evaluated in air. Both AEDA and Charm methodologies originally proposed that the larger the dilution value, the greater the potential contribution of that compound to the overall aroma. With time, data interpretation has changed. Researchers now consider AEDA, OAV and Charm methodologies to be screening in nature. These methodologies are used to determine which aroma compounds most likely make a contribution to the odor of a food recognizing that sensory work (e.g. recombination studies) needs to be done to determine which aroma compound are truly contributory. If we are to attempt any recreation of a flavor, we must have a list of odorants to study and have some reasonable basis for their selection and ranking. There are numerous references in the literature where the application of these methodologies have resulted a good recreation of food aroma (*6-9*).
   Despite the recognized limitations and successes of the GCO methods discussed, we chose to determine whether odor activity values can be used to predict the perceived intensity of odorants and whether they could predict odorant contribution to perception of a mixture. While there has been substantial criticism of these methods in the literature (*10-14*), there has been little solid data to validate or refute the criticisms. This paper presents the results of two similar sensory studies. Study 1 used vanillin, furfuryl mercaptan, furaneol and skatole as odorants; study 2 used vanillin, furfuryl mercaptan and cis-3-hexen-1-ol. In both studies, the detection thresholds of the compounds were determined. Next, the odor intensity function (relation of perceived intensity to concentration) for each compound was established. Finally, we determined the contribution of a single compound to the overall intensity of a binary mixture.

MATERIALS

Subjects

In both studies, ten subjects participated in sensory analysis. The participants were students and staff at the University of Minnesota, Department Food Science and Nutrition. None reported any problems with smell ability. They were not paid to participate in these sensory tests. In Study 1, seven women and three men, between 23 and 40 years old, served as subjects. In Study 2, five women and five men, between the age of 23 and 41 served as subjects. Two of the five women and two of the five men had participated in Study 1 (one year separated the studies).

Compounds

Five odorants were used: vanillin, Sigma Chemical Co, Saint Louis, MO (USA); furaneol, 95%, Aldrich Chemical Company Inc., Milwaukee, WI (USA); furfuryl mercaptan, 98%, Aldrich Chemical Company Inc., Milwaukee, WI (USA), cis-3-hexen-1-ol (98%), Aldrich Chemical Company Inc., Milwaukee, WI (USA), and skatole, 98%, Acros Organics, Pittsburgh, PA (USA). The compounds were chosen to represent both pleasant and unpleasant odors and were used at the purities as received. The issue of purity is addressed in the Results and Discussion sections.

Diluent

The diluent used was drinking water supplied by Kandiyohi Bottled water Co., Willmar, MN (USA). For Study 1, the stock solutions of furfuryl mercaptan ($2*10^{-4}$ g/g diluent) and skatole (10^{-4}g/g diluent) contained ethanol at 10 % and 15 % respectively. The solutions of furfuryl mercaptan contained only ethanol in the determination of its threshold in Study 1. In Study 2, all the dilutions were done in drinking water. Series of half log step were used for preparing the dilutions.

Delivery System

The dilutions were presented in 60 ml amber bottles closed by a Teflon screw cap (Qorpak, Fisher scientific). The bottles were 75 mm tall with a 34 mm opening diameter and a 44 mm body width. While the solutions were generally prepared the day of sensory evaluation, they were never prepared more than 24 hrs in advance. The subjects were told to shake the bottle before smelling.

METHODS

Detection Threshold

Detection thresholds were determined using a single staircase forced choice procedure (*15*). This procedure was found to give good reliability in the threshold determination of butanol (*16*). The concentration of the stimulus was increased following trials in which a subject failed to detect the stimulus and decreased following trials where correct detection occurred. The method started with the weakest concentration and moved to the highest concentration. A trial consisted of the presentation of two glass sniff bottles in rapid succession to the subject. One bottle contained 20 ml of a given concentration of the odorant dissolved in the diluent whereas the other contained 20 ml of the diluent alone. The subject's task was to report which of the two randomly presented bottles smelled stronger. Even if no difference was perceived between the two samples, the subject had to choose one or the other bottle. No feedback was given. The inter-trial interval was about 20s; this interval was not controlled because it seems to have no influence on the threshold value (*17*). A half log step (weight/ weight) dilution series was used. A bottle was opened a maximum of three times for the same subject. Ten bottles of each concentration were prepared. The staircase started at the lowest log concentration step and was moved upward in full log steps until correct detection occurred on five consecutive trials at a given concentration. If a mistake occurred on any trial before this time, the next higher log concentration was presented. When five consecutive correct trials occurred at a given concentration, the staircase was reversed and moved downwards in 0.5 log steps, with either one or two trials at each step (i.e. if the first trial was missed, the second one was not given, and the staircase was moved to the next higher concentration). The geometric mean of the last four of the seven staircase reversal points was used as threshold value.

In Study 1, two measurements of the threshold were made for each subject and each compound in two different sessions. In study 2, three measurements of the threshold for each subject/compound were made. The geometric mean of the individual thresholds was taken as the panel threshold value.

Data Analysis

Analysis of variance (ANOVA) was computed for each compound's threshold value in order to look at the subject and replication effects. For each study, the difference among threshold values of the odorants was assessed by computing an ANOVA with subjects, compounds and replication as factors. Finally, General Linear Model (GLM, SAS procedure) assessed if there was a significant difference between threshold values established in Study 1 and Study 2 for vanillin and furfuryl mercaptan.

Perceived Intensity versus Concentration

In a single session, the different concentrations of each compound (Table I) were rated for perceived intensity. The odorants were presented randomly and the subjects rated the perceived intensity on a 14.5 cm line scale labeled none at the left and very strong at the right. The marks from the line scales were converted to numbers between 0 and 10 using a digitizer. Before the session, we illustrated the range of intensities they would experience by having them smell the bottles with the weakest intensity (furaneol $10^{-6.5}$ w/w and vanillin $10^{-6.0}$ w/w) and the bottle with the strongest intensity (furfuryl mercaptan 10^{-6} w/w). The order of presentation of the samples was randomized for each subject. They waited two minutes between bottles to recover from adaptation. One replication of the test was performed.

Table I. Concentration of Odorant Compounds (w/w) for Rating Perceived Intensity

Study 1			
skatole	*furaneol*	*furfuryl mercaptan*	*vanillin*
$10^{-8.0}$	$10^{-6.5}$	$10^{-9.0}$	$10^{-6.0}$
$10^{-7.0}$	$10^{-6.0}$	$10^{-8.0}$	$10^{-5.5}$
$10^{-6.5}$	$10^{-5.5}$	$10^{-7.5}$	$10^{-5.0}$
$10^{-6.0}$	$10^{-5.0}$	$10^{-7.0}$	$10^{-4.5}$
$10^{-5.5}$	$10^{-4.5}$	$10^{-6.5}$	$10^{-4.0}$
$10^{-5.0}$	$10^{-4.0}$	$10^{-6.0}$	$10^{-3.5}$

Study 2		
cis-3-hexen-1-ol	*furfuryl mercaptan*	*vanillin*
$10^{-6.0}$	$10^{-9.0}$	$10^{-6.0}$
$10^{-5.5}$	$10^{-8.5}$	$10^{-5.5}$
$10^{-5.0}$	$10^{-8.0}$	$10^{-5.0}$
$10^{-4.5}$	$10^{-7.5}$	$10^{-4.5}$
$10^{-4.0}$	$10^{-7.0}$	$10^{-4.0}$
$10^{-3.5}$	$10^{-6.5}$	$10^{-3.5}$
	$10^{-6.0}$	

Data Analysis

The ability of each subject to establish a function between compound concentration and perceived intensity was assessed by fitting regression models using SAS version 6.11 (SAS institute, Cary, NC., USA). The following models were tested:
- linear models:
    Intensity vs. Log Concentration (Weber-Fechner 'law')
    Log Intensity vs. Log Concentration (Stevens 'law')
- Quadratic models:
    Intensity vs. Log Concentration and (Log Concentration)^2
    Log Intensity vs. Log Concentration and (Log Concentration)^2

Stevens and Weber-Fechner's law commonly describe the relation between perceived intensity and concentration. We chose to consider also a quadratic model to take into consideration that at a higher concentration, the perceived intensity may reach a plateau.

The "best model" was determined by examining the coefficient of determination (R^2) and the residual plots for each compound and for each subject. The models were computed over all the data points (no average was made over the replications). An F test was used to determine if the quadratic models better fit the data for each judge/compound pair. We considered that a subject was able to perceive a relation between perceived intensity and concentration when the coefficient of determination was above the value of 0.35. Only data from subjects with an $R^2 > 0.35$ was used in subsequent analysis. This will be referred to as the restricted panel.

After screening the ability of the subjects, we fit an intensity – concentration function for the restricted panel. At that time, the models were computed over the average of the repetitions for each single subject (if we have six subjects in the restricted panel, we have six data points for each level of concentration). When we observed in our regression analysis that the variance was not constant over the range of concentration tested, a weighted regression was computed (weight = 1/ variance).

Contribution of a single compound to the overall intensity of a binary mixture

Three mixtures containing furfuryl mercaptan and vanillin were prepared (Table II). The subjects (full panels) evaluated the contribution of each compound to the overall intensity of the mixture. The answers were given as a percentage of the overall intensity. For both studies, the samples were given in duplicate within a session. The panels participated in two sessions.

Table II. Binary Mixture Compositions

Mixture	Concentration of furfuryl mercaptan (w/w)	Concentration of vanillin (w/w)
1	$10^{-9.0}$	$10^{-4.5}$
2	$10^{-8.5}$	$10^{-4.5}$
3	$10^{-8.0}$	$10^{-4.5}$

RESULTS

Detection thresholds

The threshold values of our aroma compounds are reported in Table III. The wide variation in the sensory thresholds for each compound is notable and may be due to two factors. The first is that these data were obtained from several individuals over a period of more than one year. Large differences within and between individuals in sensitivity to a wide array of chemicals has been reported in the literature (*17-19*). Second, the mixtures for sensory analysis had to be prepared in advance of the testing (up to 24 hrs) due to the logistics of sample preparation. As is discussed in the Discussion section, some aroma degradation may have occurred during this holding period.

Table III. Detection Threshold Values for the Sensory Panels

Compounds	Geometric mean* of the detection threshold values (w/w) for the panel	Range of the geometric mean of the threshold values for the subjects (w/w)
Vanillin (Study 1)	$10^{-6.41}$ (389 ppb)	$10^{-7.688} - 10^{-5.188}$ (20.51 – 6,486 ppb)
Vanillin (Study 2)	$10^{-6.03}$ (927 ppb)	$10^{-6.83} - 10^{-4}$ (147.9- 10,000 ppb)
furfuryl mercaptan (Study 1)	$10^{-10.55}$ (0.028 ppb)	$10^{-14.25} - 10^{-9.375}$ ($5.6 * 10^{-6}$ - 0.42 ppb)
furfuryl mercaptan (study 2)	$10^{-10.325}$ (0.047 ppb)	$10^{-11.375} - 10^{-8.96}$ (0.0042- 1.1 ppb)
Furaneol (study 1)	$10^{-7.75}$ (17.8 ppb)	$10^{-10.375} - 10^{-6.31}$ (0.042- 489.8 ppb)
Skatole (study 1)	$10^{-9.806}$ (0.156 ppb)	$10^{-11.06} - 10^{-7.25}$ (0.0087 - 56.23 ppb)
cis-3-hexen-1-ol (study 2)	$10^{-7.295}$ (50.70 ppb)	$10^{-8.83} - 10^{-6.125}$ (1.48 - 749.9 ppb)

* Since the compounds were used without purification, the absolute threshold values reported in the table above should not be considered accurate.

Perceived Intensity versus Concentration

For most of the subjects, the Weber's model (Intensity vs. Log Concentration) gave a higher R^2 than the Stevens' model (Log Intensity vs. Concentration) and showed less heteroscedascity than the Stevens' model. Thus we used only the semi-logarithmic model for establishing the odor intensity functions. Some judges showed a low R^2 regardless of the model applied to the data. The judges who gave inconsistent results ($R^2 < 0.35$) were removed from the final data set. In Study 1, one subject's data for furaneol and skatole was inconsistent; three other subjects' data for vanillin were inconsistent. In Study 2, two subjects were inconsistent for vanillin and another for cis-3-hexen-1-ol. Thus, six judges and seven judges were kept, respectively, in Study 1 and Study 2 for establishing the odor intensity functions of the different compounds.

Figures 1 and 2 present the odor intensity functions obtained in Study 1 and Study 2. In both studies, the slope of the intensity vs. concentration function was strongly dependent on the aroma compound. For instance, the slope of the odor intensity functions of furfuryl mercaptan and skatole are much steeper than those observed for vanillin and furaneol (Study 1).

OAV and Perceived Intensity

OAVs did not accurately predict the relative odor intensities of different odorants. In both studies the OAVs were calculated using the geometric mean of the threshold values of the subjects in the selected panel. In Study 1, the predicted intensities for the four compounds were calculated for an OAV of 2,000 (Figure 3). For an OAV equal to 2,000, the predicted intensities of vanillin, furfuryl mercaptan, furaneol, skatole were respectively 4.2, 4.3, 5.3 and 7.0. Ideally, these compounds would have the same intensities at the same OAV. However, the slopes of the plots are clearly different and going up the plot a factor of 2000 times the detection threshold results in a different intensity for each compound. Conversely, we calculated the OAVs of the compounds for two different perceived intensities (Figure 4). In Study 1, for a perceived intensity of 4, the OAVs of vanillin, furfuryl mercaptan, furaneol, skatole were respectively 1663, 1586, 724, and 219. In study 2, for a perceived intensity of 5, the OAVs of cis-3-hexen-1-ol, vanillin, furfuryl mercaptan, were respectively 662, 324, and 209.

Similarly, since the slope of the odor intensity function depends on the odorant, the rank order of the odorants according to their OAVs at a specific intensity did not predict the OAVs' rank order of those odorants at a lower or higher intensity. In Study 2, for a perceived intensity of 5, furfuryl mercaptan had the lowest OAV = 209 in comparison to the two other compounds (OAV of cis-3-hexen-1-ol = 662; OAV of vanillin = 324), whereas for a perceived intensity of 2, vanillin had the lowest OAV = 24 in comparison of cis-3-hexen-1-ol (OAV = 48) and furfuryl mercaptan (OAV = 35). Thus, while OAVs are logically used in selecting potential key odorants in a food, there is danger in ranking the importance of odorants based solely on OAV. The habit of listing odorants in tables in order of OAV unintentionally implies a ranking and it must be recognized that ranking is not intended or necessarily valid. To the credit of OAV, one should note that the OAV of the compounds in Study 2 at a low intensity are relatively close. It is generally accepted that OAV differing by less than a factor of 2 are not importantly different. Our data at lower intensities support this.

Contribution of a Single Compound to the Overall Intensity of Binary Mixtures

The entire panel averages of the percent contribution of each odorant to binary mixtures are shown in Figure 5. As is shown in this figure, an increase in the concentration of furfuryl mercaptan in a vanillin/furfuryl mercaptan mixture resulted in a decrease in the perceived contribution of vanillin. In this part of the work, the concentration of vanillin was kept constant while the concentration of furfuryl mercaptan

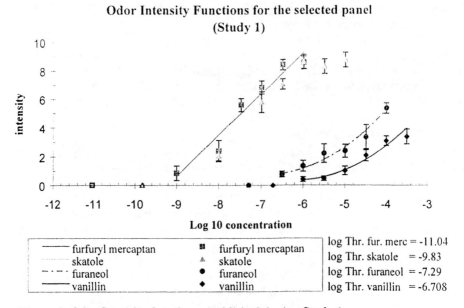

Figure 1: Odor Intensity functions established during Study 1.
The perceived intensity, detection threshold values and regression lines are the average values for the restricted panel. The error bars represent the standard error of the means.

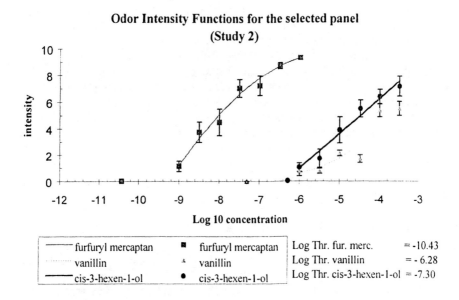

Figure 2: Odor Intensity functions established during Study 2.
The perceived intensity, detection threshold values and regression lines are the average values for the restricted panel. The error bars represent the standard error of the means.

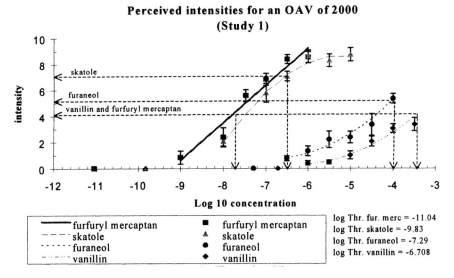

Figure 3: Perceived Intensities for the same Odor Activity Value.

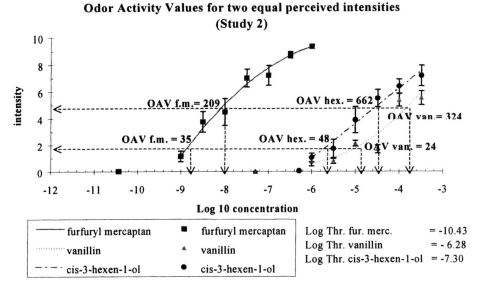

Figure 4: Odor Activity values for the same Perceived Intensity.

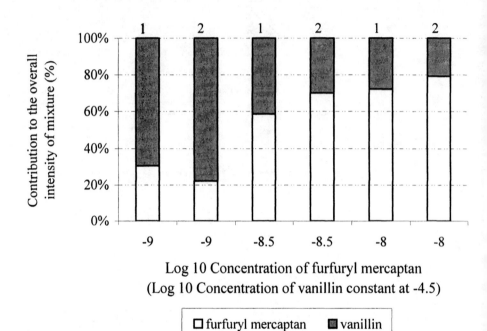

Figure 5: Contributions of a single compound to the overall intensity of mixtures. The results from Study 1 are indicated by the number 1 whereas the results from Study 2 are indicated by the number 2.

was increased. The sensory panel was asked to estimate the contribution of vanillin and furfuryl mercaptan to each mixture. Our goal was to determine if OAVs could be used to estimate the relative contribution of a given odorant to a mixture. While one might predict that this would not be possible based on the results described above, this first work considered only intensities of a single odorant, not the contribution of an odorant to a mixture. While current researchers have not proposed that OAV can be used in an absolute sense to predict the contribution of an odorant to a mixture, there has not been any data to validate or refute the hypothesis proposed by early workers that OAVs are additive and that the contribution of an odorant is equal to its OAV/sum of all OAV (*1, 2*).

The OAVs also failed to predict the relative contribution of the two odorants to their perceived *intensity* in the mixture. The correlation between the OAV ratio and the perceived sensory contribution ratio (Figure 6) is very low ($R^2 \cong 0$ for study 1 and $R^2 = 0.15$ for Study 2). If the concept of OAV were able to predict the contribution of an odorant to the overall intensity of a mixture, the data points should be close to the line y = x and the R^2 much higher. For the three mixtures studied, individual ratios (OAV furfuryl mercaptan/OAV vanillin) were computed based on the data obtained in the first part of our experiments. As there was a large variability in the threshold values between subjects, these ratios were expected to reflect the same variation. Similarly, from the results obtained in our last experiment, individual ratios of % contribution due to furfuryl mercaptan and % contribution due to vanillin were computed for the three mixtures. The ratios accurately represented the perceived contribution of an odorant to the overall intensity of a mixture. Unlike the ratios of OAVs, the measured sensory contribution showed much less variability between subjects.

DISCUSSION

As is noted in the Materials section, aroma compounds were used at the purities received. There was some concern that impurities in the aroma compounds may confound the results of this work. While it is possible that the absolute thresholds were influenced by impurities, purity is not relevant to the general principles being tested. Even if we were collecting data on the impurities as opposed to the named substances, the results pertaining to whether OAV is a useful predictor of perceived intensity or overall contribution to a mixture would be unchanged. Albeit we would have different absolute sensory thresholds since the impurities would be at an unknown concentration but the results would be the same. The same "impure" compounds were used in determining sensory thresholds, OAV and in determining the contribution of the named compound in the mixture studies.

Odor detection thresholds are difficult values to measure. Several procedures have been developed to estimate their values (*20,21*). One must be aware of the strong dependence of the threshold values on the procedure used to obtain them (*21*). Generally, these procedures are tedious and time consuming for both experimenters and subjects. Even with well established procedures to estimate the odor detection threshold values, studies have demonstrated a large variability in sensory thresholds across sessions, and

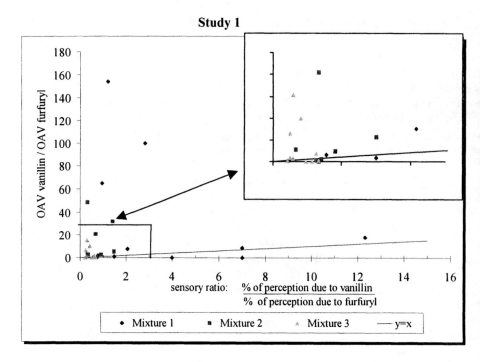

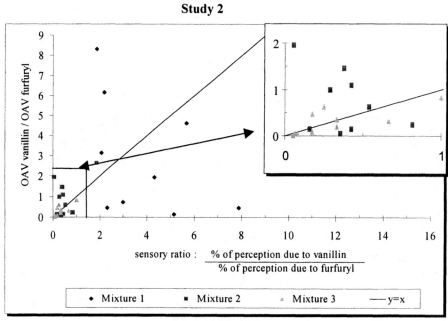

Figure 6: Plot of the ratio of the OAVs against the ratio of the contributions for both studies. Each dot represents the average values for one subject.

among and within subjects for a single compound. It is quite common to observe several log units of variability within a group of subjects (*19*). Stevens et al. (*19*) noticed a large variability within a given individual but pointed out that the mean threshold was uniform across the three studied subjects. In the same paper, Stevens reported that many factors may affect the determination of threshold values: physiological (age, gender...) as well as methodological (delivery system, experimental procedure...). Since OAVs are dependent upon sensory thresholds, rigorous techniques should be used in their estimation and the variability inherent in their determination should be considered in using OAVs as absolute numbers.

Some features of our methodology may have added variability to our threshold data. We used small sniff bottles with a limited headspace volume. This could have affected our threshold results i.e. raised the values. Small headspace volumes provide less odorants to the olfactory epithelium (*17*). However, as the same bottles were used throughout our studies, the choice of this delivery system should not have influenced our relative conclusions. A second source of variability could have been the lack of consistency of concentration in the gaseous phase. However, we had subjects shake the bottle each time before smelling the sample. So, we assumed constant concentration in the headspace. A third problem may have been the length of time between sample preparation and sensory testing. Samples were occasionally prepared as much as 1 day before sensory analysis due to the time required for their preparation. Unstable molecules such as furfuryl mercaptan likely underwent some degradation during storage resulting in added variation in thresholds (fresh one time but stored another). Molecules such as vanillin are stable in aqueous systems and exhibited relatively low variability in sensory thresholds (300X). Furfuryl mercaptan is relatively unstable and its sensory thresholds were found to vary by a factor of nearly 10,000 fold. One should note that despite the wide range in sensory thresholds found in this research, the geometric averages between our two studies were very close.

OAV was a poor estimate of odor intensity across compounds at higher intensities but a reasonably good estimate of intensities at lower intensities. Because the slopes of the odor intensity functions are not equal for all the odorants (*22*), two components having equal OAVs would not be expected to exhibit the same perceived intensity. Also, OAV was not a good predictor of the perceived contribution of an odorant to its overall intensity in a binary mixture. Since GCO techniques in general do not consider interactions between odorants, one would again not expect OAV to be an accurate indicator of the contribution of an odorant to a mixture. Thus, we were not surprised at the poor relationship between OAV and % contribution of our odorants to a mixture. There have been few studies on odor mixtures in the sensory literature. Several models have been developed to predict the perceived intensities of binary or ternary mixtures (*23*) but far less is known for more complex mixtures (*24*).

In summary, AEDA, Charm and OAV data should be considered solely as screening techniques to select aroma compounds for further sensory studies. While the presentation of data in tables listing odorants in order of OAV, FD or Charm value is done out of custom (not unlike the listing of odorants in order of elution from a GC), the reader must be cautioned that this listing does not necessarily imply that the odorants are truly listed in order of importance or contribution to overall odor. The true contribution of an odorant to overall aroma can only be determined via rigorous sensory testing. It is our opinion that while these GCO techniques have limitations, they represent the best screening methods available in our field.

The use of sensory testing for validation of aroma contribution must be done carefully. One approach currently being used is omission testing (compare the full odorant recombination to the recombination minus one odorant). However, one must recognize that an odorant may make a contribution to overall odor but its contribution may be small and be lost in the variability of the method. Thus, one may eliminate odorants that make small but significant contributions to odor quality. Also, researchers must use sensory techniques designed to evaluate similarity in their final evaluation (21) i.e. is the recombination close to the target sample and exactly how close? One could use triangle testing to determine if the recombination matches the real product. Unfortunately, other clues (e.g. the texture of a model cheese vs. a real cheese) may give sample identity away and invalidate this approach. Attribute testing can be used to give an ideal of similarity, but it is not a valid method for estimating overall flavor match. The fields of flavor chemistry and sensory science must continue to merge for success in this endeavor.

Literature Cited

1. Patton, S.; Josephson, D. *Food Res.* 1957, 22, 316- 318.
2. Guadagni, D.G.; Buttery, R.G., Harris, *J. Sci. Food. Agric. Technol.* 1966, 17, 142-144.
3. Ullrich, F.; Grosch, W. *Z. Lebensm. Unters. Forsch.* 1987, 184, 277-282.
4. Kerscher, R.; Grosch, W. *Z. Lebensm. Unters. Forsch.* 1997, 204, 3.
5. Acree, T.; Barnard, J.; Cunningham, D. *Food Chem.* 1984, 14, 273- 286.
6. Buttery, R.G.; Teranishi, R.; Ling, L.C. *J. Agric. Food Chem.* 1987, 35, 540.
7. Guth, H.; Grosch, W. *Amer. Chem. Soc.* 1993, 70(5), 513-518.
8. Zehentbauer, G.; Grosch, W. *J. Cereal Sci.* 1998, 28 (1), 81-92.
9. Schieberle, P.; Hoffman, T. *J. Agric. Food Chem.* 1997, 45 (1),
10. Rothe M. *Die Nahrung.* 1976, 20 (3), 259- 266.
11. Frijters, J.E. *Chem. Senses Flavour.* 1978, 3 (2), 227- 233.
12. Piggot, J.R. *J. Sensory Studies.* 1990, 4, 261- 272.
13. Abbot, N.; Etievant, P.; Issanchou, S.; Langlois, D. *J. Agric. Food Chem.* 1993, 41, 1698 - 1703.
14. Mistry, B.S.; Reineccius, T.; Olson, L.K. In *Techniques for Analysing Food Aroma*; Marsili, R.T., Ed.; Dekker, M.: New York, NY, 1997; pp 265 - 292.
15. Doty, R. L. In *Handbook of Olfaction and Gustation*; Doty, R.L., Ed.; Marcel Dekker, New York, 1995, pp 191 – 225.
16. Doty, R.L.; McKeown, D. A.; Lee, W.W.; Shaman P. *Chem. Senses.* 1995, 20, 645-656.
17. Doty, R. L.; Gregor, T.,P.; Settle, R. G. *Chem. Senses.* 1986, 11 (2), 259 - 264.
18. Friedrich, J.; Acree, T.E. Amer. Chem. Soc. Meeting Abstracts. #39. August 1999.
19. Stevens, J.; Cain, W., Burke, R. *Chemical Senses.* 1988, 13, 4, 643- 653.
20. Bi, J.; Ennis, D.M. *J. Sensory Studies.* 1998, 1, 133- 148.

21. Lawless H.; Heymann, H. In *Sensory Evaluation of Food: Principles and Practices*. Heldman, D.R., Eds; Chapman & Hall: New York, USA, 1997. Chapter 6, pp 173 – 207.
22. Patte, F.; Etcheto, M.; Laffort, P. *Chemical Senses and Flavor*. 1975, 1, 283- 305.
23. Laffort, P. In *Perception of complex smells and tastes*. Laing, D.G., Cain, W.S., McBride R.L., Che, B.W. Eds. Academic Press, New York, USA, 1989, pp 173 – 188.
24. Laing, D.G. In *Handbook of Olfaction and Gustation*; Doty, R.L., Ed.; Marcel Dekker, New York, 1995, pp 285.

Chapter 15

Odor Intensity Measurements in Gas Chromatography–Olfactometry Using Cross Modality Matching: Evaluation of Training Effects

G. Callement, M. Bouchet, D. Langlois, P. Etiévant, and C. Salles

Laboratoire de Recherches sur les Arômas, INRA, 17 rue Sully, 21034 Dijon Cédex, France

The aim of this paper was to report the GC-O training effect on the performances of a panel evaluating odor intensities in gas chromatography effluents by cross modality matching with the finger span method. The panel was first trained with different concentrations of ethylbutyrate and, in order to test the influence of the odor quality on the performance of each panelist, ethylbutyrate and hexanal were then alternatively used. Training was followed by the evaluation of model mixtures containing 11 different compounds in various concentrations. An analysis of beer samples was finally performed by trained and untrained panelists to test the effect of training on panelist performance ; that was a larger number of odors detected, a better estimation of the odor intensity leading to a better discrimination of the samples.

Several detection techniques have been coupled to gas chromatography to analyze the odors from foods, perfumes,... Among them, GC-O is particularly interesting because of the high sensitivity of the human nose which is often higher than any electronic detector. Acree et al. (1) and Ullrich and Grosch (2) were the most important contributors in the development of this technique by the introduction of two variant methods : CHARM and AEDA respectively. Since this time, a large number of studies were achieved and reviewed by Grosch (3) and Mistry et al. (4). However, these methods present some limitations in the evaluation of odor intensity : firstly, the important number of dilutions of the extract which have to be tested is time-consuming, leading to take only a few number of panelists (or just one), and secondly, the results are based on detection thresholds and not on real odor intensities. The OSME method (5 ; 6) overcomes these difficulties because the panelists evaluate the odor intensities directly by a magnitude estimation with a variable resistor equipped with a pointer moving along a scale. This system was proved performent because the authors obtained a good correlation between the estimation of odor intensity and the concentration of volatile compounds. Later, Etiévant *et al. (7)* used successfully a variant evaluation method : the cross-modality matching with finger span. The use of synthetic solutions showed that panelists were able to determine most of its characteristics with a very high correlation between the concentration of the components and their finger span. The aim of this paper is to test the ability of a panel composed of trained and untrained members to evaluate the odor intensity with simple synthetic samples and to discriminate complex synthetic mixtures and beer aroma extracts using the finger span cross-modality matching. We will pay special attention to the effect of panelist training.

Materials and Methods

Chemicals. Aroma compounds were purchased from Sigma-Aldrich (St Quentin Fallavier, France). Dichloromethane (pure for synthesis, SDS, Peypin, France) used as dilution solvent was freshly distillated before use. The 4 beer samples were given by TEPRAL (Danone, Strasbourg, France).

Solutions. The synthetic solutions were prepared taking into account the Stevens parameters to calculate the concentration series for each aroma constituent *(7)*. The beer aroma extracts were obtained as described by Cunnigham *et al. (8)*.

Gas chromatography. The analyses were performed on a HP 5890 instrument equipped with a split-splitless injector (210°C ; split ratio 1 : 3.3) and a sniffing port. The column, DB1701 (J&W Scientific Inc ; 15 m ; 0.53 mm i.d. ; 1 μm thickness), was connected to the FID and the sniffing port with a Y press fit connector equipped with capillaries of equal lengths. The carrier gas was hydrogen (57 cm.s^{-1}). The transfer line to the sniffing port was heated to 250°C. Humid air (100 ml.min^{-1}) was added to the chromatographic effluent at the bottom of the glass sniffing cone.

For ethyl butyrate, 1 μl of each of the six different solutions was successively injected every 30 sec with a different and labeled syringe. The split was opened 330 sec after the first injection. The temperature of the column was maintained at 40°C during each analysis.

For the synthetic solutions, 1 μl of each one was injected on the column maintained at 40°C for 12 min. The injector split was opened after 30 sec. Afterwards, the oven temperature was raised to 215°C at 6°C.min^{-1} then to 220°C at 10°C.min^{-1}.

For beer aroma extracts, the analyses were performed in the same conditions and on the same instrument as described above, but equipped with an on-column injector and a DBWAX column (J&W Scientific Inc ; 30 m ; 0.32 mm i.d. ; 0.5 μm thickness). A silica capillary without active phase (J&W Scientific Inc ; 0.7 m ; 0.32 mm i.d.), fitted between the injector and the column, was changed every 2 beer extract injections and several runs without extract injection were necessary to avoid interference.

Gas chromatography – Olfactometry. The panelists were instructed to match the intensity of the perceived odors during the analyses with the finger span device as described by Etiévant *et al. (7)*. The evaluations were realized in a quiet and isolated room at 21°C.

Panelists. The 7 panelists were volunteers and coded as OC, CA, LB, FM, CD, MH and MJV. CD stopped the evaluations after the butyrate experiment. CD, MH, LB and MJV participated regularly to descriptive sensory evaluations in the laboratory and moreover, LB and MJV had previous experience in GC-O.

Familiarization sessions. During 3 sessions of familiarization, a series of 6 ethyl butyrate solutions of decreasing concentrations (13350, 7870, 4110, 1770, 530, 67 in mg.l^{-1}) corresponding respectively to the theoretic olfactive intensity calculated from Stevens parameters (27.8, 23.1, 18.4, 13.7, 9.0, 4.3) were injected successively. This series of stimulation was repeated 4 times in each session. The panelist, informed of the decreasing intensity of the stimulus within each series, was asked to match the different intensities perceived with the finger span sensation.

Training sessions. Four series of 6 stimulations with different concentrations of ethyl butyrate (see above) were presented to the panelists, firstly in a decreasing concentration and afterwards in disorder, the first concentration being always the highest. The panelists were informed of the order of the concentrations and in order to make a self calibration of the prototype, they were asked to fit the perception of the first stimulus with their maximum finger span. Three to nine training sessions were necessary according to the discrimination ability of each panelist evaluated by a variance analysis. In order to minimize adaptation, the time lag between two stimulations was settled to 60 sec.

Evaluation sessions with ethyl butyrate. The evaluation sessions were identical to the training sessions. Only 5 panelists participated to the evaluation sessions (MH and CD were discarded after the training sessions). An experimental design balanced on six sessions and five panelists for order and first order carry-over effects was used. The panelists were not informed of the order of presentation of sample concentration apart from the first one corresponding to the highest concentration.

Evaluation sessions with ethyl butyrate and hexanal. The procedure previously described was modified as follows. Ethyl butyrate and hexanal were alternatively injected at different concentrations according to an experimental design balanced on four to six sessions. Each session consisted in 4 series of 6 stimulations. The first one corresponding to the highest concentration (reference) was alternatively ethyl butyrate or hexanal. The time between 2 consecutive stimuli also varied according to the series, from 21 sec to 149 sec. For ethyl butyrate, the concentrations used were the same as described above in Familiarization session. The concentrations of hexanal were: 7995, 4848, 2620, 1180, 379, 53, (in mg.l^{-1}) corresponding respectively to the theoretic olfactive intensity : 27.8, 23.1, 18.4, 13.7, 9.0, 4.3.

Evaluation sessions with synthetic solutions. Each panelist evaluated four solutions containing eleven volatile compounds present in different concentrations (Table I) for six sessions. Two different solutions amongst four were analyzed in each session. For each sample, the panelists were informed of the temperature of elution of each volatile component and they were allowed to check the temperature of the oven during the analysis.

Evaluation sessions with beer aroma extracts. Each panelist evaluated four beer samples with three repetitions in the same conditions as described above for the synthetic solutions. The panelists OC, CA, LB and MH, who were subjected to familiarization, training and measure sessions on simple and complex mixtures, constituted the trained panel. The panelists MB, FC, CB and YL who were only subjected to familiarization session constituted the untrained panel.

Experimental designs were obtained with the FIZZ system from Biosystème, France. Univariate statistical analyses were realized using the Sigmastat scientific software (version 2.0) from Jandel Corporation (Erkrath, germany). Multivariate statistical

Table I. Synthetic Solutions: Constituent Concentrations (Gram per Liter) and Corresponding Odor Intensity Range

Compound	Solution n°				T.I.R.*
	1	2	3	4	
3-Methylbutane-thiol (MBT)	0.002	0.514	0.238	0.088	1-6.2
Hexan-2-one	10	15	10	15	1-1.2
Furfural	15	5.1	15	5.1	1-1.7
Benzaldehyde	9	15	9	15	1-1.2
Octan-1-ol	7	7	15	15	1-1.2
Nonanal	15	5.21	0.4	15	1-2.7
2-Methoxyphenol (guaiacol)	0.486	1.61	4.09	0.09	1-6.0
Citronellal	15	1	9.36	15	1-2.3
2-Phenyl-1-ethanol	0.112	2.16	3.88	0.42	1-6.1
Decanal	15	0.6	5.61	5.6	1-2.7
Vanillin	0.071	0.273	1.5	0.7	1-5.9

* Theoretical Intensity Range calculated from the theoretical intensity of the concentration corresponding to an odor just detectable (Reproduced from Reference 7. Copyright 1999 American Chemical Society).

analyses were made with the Statbox plus software (version 2.0) from Grimmer logiciels (Paris, France).

Results and Discussion

The cross modality evaluation method used to estimate the odor intensity detected by GC-O does not require a feed back information to the panelists (9) and is rather different than the OSME method described by McDaniel et al. (5) and Da Silva et al. (6). The aromagramme obtained represents a succession of positive peaks where only the maximum height of each peak is considered because panelists were instructed to evaluate the intensity of each odor with their finger span without taking in account the time factor.

After each session, the evaluation of the performance of each panelist was made from an analysis of variance with concentration as the only factor, on the basis of the peak height values. For each panelist, we observed an increase of the peak heights with the concentration (H° rejected at $p = 0.05$). A Newman-Keuls test was performed to determine the number of pair of consecutive concentration (CPC) significantly different at $p = 0.05$. As the number of different stimulus concentrations was six, the maximum score was five with, theoretically, the same difference of intensity between each consecutive pair. A large difference in the performance of the panelists was observed. As they were informed of the order of presentation of the concentrations, the panelists

FM and MJV were discarded because they were not able to discriminate more than 3 CPC (data not shown), that was a too low performance. However, the other panelists were able to discriminate an average of 4 CPC. The figure 1 shows the evolution of the performances of each panelist for each session. Depending on the initial performance of the panelists and on their progresses, the number of sessions varied from one panelist to the other. On this figure, we can observe neither a clear effect of the training on the performances of the panel, nor clear differences between the first sessions where the stimuli were presented in a decreasing order and the last sessions where the stimuli were presented without special order. Most of the panelists present irregular performances but the average performances of someone's (CA, LB and MH) are significantly higher than the others (particularly for FM and MJV, p=0.05).

For the measure sessions, the only difference to the training sessions is that the panelists were not informed of the order of presentation of the stimuli. Figure 2 shows that the performances of the panelists does not decrease systematically if they are not informed of the order of presentation of the stimuli. The performances of OC and LB increase while the performances of CA, CD and MH decrease.

To add some difficulty, alternated injections of ethyl butyrate and hexanal were made. As presented in Figure 2, we did not notice any effect compared to the performances recorded with stimulation with only ethyl butyrate. The difference of the performances between the panelists seemed increased with the difficulty of the test (i.e. significant difference between OC and MH, and between LB and MH). Figure 3 represents the evolution of the performances of the panelists for the measure sessions for ethyl butyrate only (sessions 1 to 6) and ethyl butyrate/hexanal (sessions 7 to 12).

We could notice in particular some differences of performances between panelists. These performances were irregular and were not improved with the number of measures. For CD, we could observe a decrease in the performances as the sessions were going by, that being probably due to weariness. Though the training sessions were conducted with ethyl butyrate, the measure of the intensity of hexanal seemed similarly good because no significant differences were observed between the performances (p>0.05).

As our main objective was to evaluate the odor intensity of food flavor extract, more complex synthetic solutions were prepared and evaluated. Eleven substances were chosen for their different volatility and chemical functionality, their known Steven's exponent (10) and because they were detected by the panelists at a concentration lower than the GC saturation concentration. The concentrations and the intensity of stimulation of individual compounds in the four solutions are reported in Table I.

Figure 4 represents an analysis of variance on the factor solution for each stimulus to evaluate the performance of each panelist. We can mainly notice the low individual performances of the panelists. In the case of this study, related to Table I, the low theoretical intensity range cannot explain that the different concentrations of hexan-2-one and octan-1-ol were found different while compounds such as guaiacol and 2-phenyl-ethanol were not found different with a high theoretical intensity range. In fact, each descriptor is discriminant between samples for a few panelists and only the cumulative information leads to an efficient discrimination of the products. These low individual performances show the necessity to use several panelists to perform this kind of analysis.

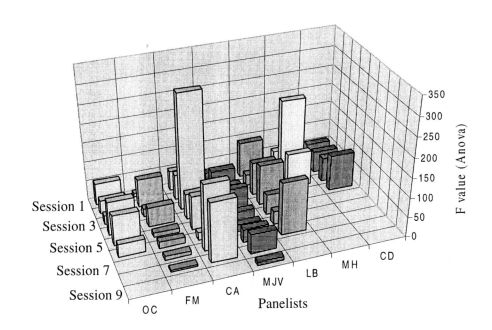

Figure 1. Training with ethyl butyrate : evolution of the performances (p<0.05).

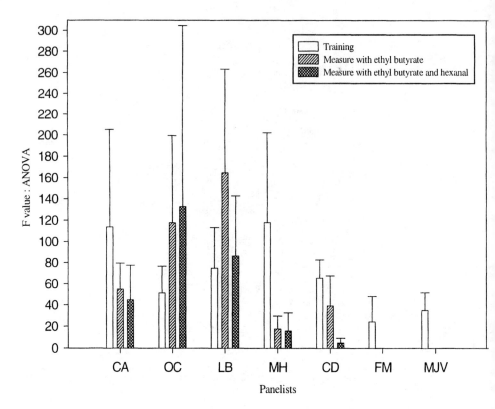

Figure 2. Training and measure : comparison of the performances of the panelists with ethyl butyrate (F value, ANOVA).

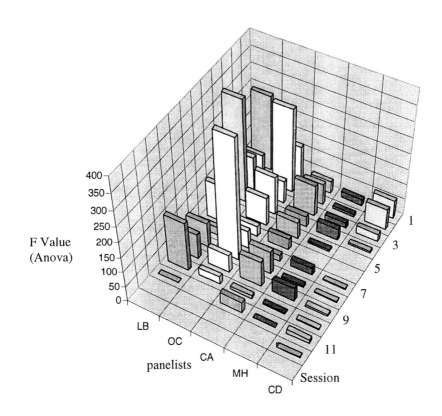

Figure 3. Measure with ethyl butyrate and hexanal : evolution of the performances.

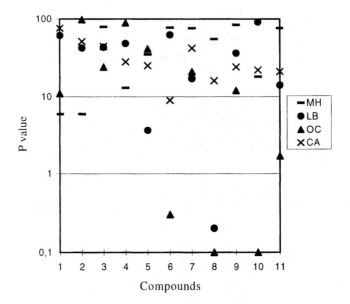

Figure 4. Evaluation of the synthetic solution: ANOVA on the concentration factor per panelist and per variable (p<0.05) (Reproduced from reference 7).
(1: MBT; 2: hexan-2-one; 3: furfural; 4: benzaldehyde; 5: octan-1-ol; 6: nonanal; 7: guaiacol; 8: citronellal; 9: phenylethanol; 10: decanal; 11: vanillin).

The agreement of the panel was evaluated by a two-way analysis of variance presented in Table II, taking into account the factors panelist and solution with interactions, with panelists as a random factor. We can observe in particular an important solution effect.

A PCA (not represented) made from the raw data (means of the repetitions for each panelist for each solution) mainly shows the panelist effect noticed on table II. To eliminate this effect, the data were centered for each panelist and each compound (subtraction of the general mean from each of the raw data obtained). The PCA is presented on Figure 5. The plot formed by the two first principal components explains 56% of the total variance and shows clearly the solution effect. The first principal component discriminates clearly the solutions 1 and 4 from the solutions 2 and 3 while the second principal component discriminates rather the solutions 1 and 2 from the solutions 3 and 4. Otherwise, we noticed that the observations related to solutions 2, 3 and 4 are well grouped while those provided from the solutions 1 are more dispersed with, in particular the point CAS1 showing that this panelist perceived the solution 1 very different from the solutions 2 and 3, as the other panelists, but that he was unable to discriminate the solution 1 from the solution 4.

Table II. Two-Factor ANOVA of the Peak Heights Obtained from GC-O-FSCM of the Synthetic Solutions

	F values		
Variables	*Panelist effect*	*Solution effect*	*Interaction effect*
3-Methylbutane-thiol (MBT)	8***	19.9***	0.8
Hexan-2-one	3.3*	10***	0.9
Furfural	1.7	14.9***	0.7
Benzaldehyde	0.5	15.7***	1.2
Octan-1-ol	4.9**	5**	1.6
Nonanal	7.8***	9.6***	1.4
2-Methoxyphenol (guaiacol)	0.9	70***	1.8
Citronellal	17.4***	12.3***	1.3
2-Phenyl-1-ethanol	2.9*	18.3***	0.4
Decanal	5.3**	6.7***	0.8
Vanillin	4.4*	67.4***	1.7

***, $p < 0.001$; **, $p < 0.01$; *, $p < 0.1$ (Reproduced from Reference 7. Copyright 1999 American Chemical Society) .

To evaluate the relation between the theoretical and the estimated intensities, the logarithm of the individual mean peak heights obtained from the evaluation of each solution were regressed on the logarithm of the corresponding theoretical intensities (log „ perceived intensity " = n log „ finger span peak height ") *(7)*. Eight variables among eleven show a significant log-log relation ($p<0.05$) between the two cross-matched sensory modalities. That shows that the panel was able to evaluate correctly 8 of the 11 presented compounds but it was unable to do that for hexan-2-one,

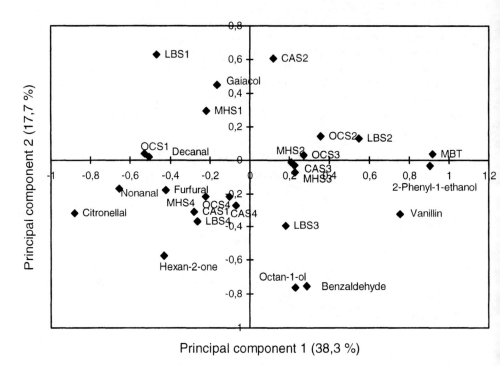

Principal component 1 (38,3 %)

Figure 5. PCA representation of the synthetic solutions (S1-S4) along the two first components for each panelist.

benzaldehyde and guaiacol which were found with a very low correlation between these two variables. Nevertheless, some different concentrations are not clearly distinguished by the panelists for the eight well correlated compounds and are totally confused for the three non correlated compounds. These observations cannot be related directly to the different range intensity of the stimulus from one compound to another (Table I). An explanation of this observation is that the Stevens exponents used to determine the theoretical intensities for these three compounds were much lower than the exponents corresponding to our panelists.

After these rather encouraging results, we decided to start with food products. The study concerned four beers with three evaluations by beer, performed by the four trained panelists who participated to the previous studies (OC, LB, CA, MH) and by four untrained panelists who never used the prototype (YL, FC, MB, CB). The three first untrained panelists were used to olfactometry analyses but the last one was a total novice. One hundred and fifty seven odors were detected by the eight panelists with a great number of very various descriptors. Only ten of them were detected by all the panelists and fifty six odors were detected by at least four panelists.

The total number of detected odors for each panelist varies from 57 to 141 according to the sensitivity of each one. However, we noticed that the total number of detected odors is more important for the trained panelists than the untrained ones. It can be explained by a better sensitivity of the trained panel or by a better positioning of their nose towards the fragrance outflow.

To determine the ability of the panelists to evaluate the perceived odors with the FSCM prototype, an analysis of variance (1 factor : beer) was made for each panelist on the variables for which the detection was observed at least twice by panelist for at least one beer. The number of significant variables was rather important (data not shown) but their comparison between panelists is difficult to interpret because the variables (odorant zones) are often different from one subject to another (OC was discarded because he tested only two beers) and the intensity of some odorant zones varies according to the beer extracts. However, a training effect is significantly observed (Ki^2 test, $p < 0.05$) on the number of discriminant variables.

The comparison between the panelists on the fifty six variables detected at least by four panelists led to the data reported on Figure 6. The panelists MB, MH and FC use the prototype with highest efficiency while the performances of LB, CA, and CB are medium and that of OC and YL are the lowest. In this case, we did not observe the apparent training effect noted above. So, the training did not improve the performances of the panelists for the measure of the intensity with the FSCM prototype and two familiarization sessions seemed sufficient for an optimal utilization of the FSCM prototype.

Another important aspect to evaluate is the ability of the panelists to modulate the intensity of odors with the prototype. To evaluate this ability, two analyses of variances were made to determine the number of significant variables (not shown), one taking into account the nil variables corresponding theoretically to the non perception of odor, and the other after discarding these nil variables. The decrease of the percentage of significant variables observed after comparison of these two analyses led to the estimation of the capacity of modulation of the signal for each panelist (Figure 7).

This comparison shows that all the panelists were able to modulate the perceived odors on their finger span with a relatively good efficiency because the number of significant variables is always different from zero and because the decrease of this number are lower than 25 % in the worse case (MH) when the nil values are discarded. However, we must indicate and take into account that this particular panelist suffered of metacarpal problems leading to limited performances. Regarding the training aspect, it is clear that trained panelists OC, LB, and CA were more able to use the finger span to modulate the odor intensities than untrained panelists. Moreover, we noticed that the increasing order of performances for the trained panelists was rather similar to the observations made for the familiarization measures with ethylbutyrate. So, we think that this compound should be very useful for selection tests.

Global analyses with all the panelists were made on the 56 selected variables which are common at least for four panelists (results not shown). As each serial of beer was repeated three times, an analysis of variance was made on the repetition factor to evaluate the evolution of the performances of the panel. A repetition effect (Kruskal-Wallis test, $p<0.1$) was observed only for three variables but the evolution of the means for each repetition did not show a particular tendency for these variables. As we previously noticed, panelists training did not seem to have some effects on their performances. An analysis of variance with two factors (Panelist, beer and interaction panelist*beer) shows an important panelist effect for most of the variables due to variations in the manipulation of the prototype. A beer effect was also observed for 22 of the 56 variables, showing that panelists were able to discriminate the beers with the prototype. Nevertheless, 15 of these variables presented a significant interaction which could be due to a different relative notation of beers according to the panelists. Then, the test of the beer effect on the interaction led us to discard 4 of these 15 variables.

To conclude this study, the training of the panel seems mainly to increase the number of odors detected and to improve the ability of the panelists to modulate the signal with the FSCM prototype. "CHARM" or "AEDA" analysis request a high number of dilutions per aroma extract. That is very time-consuming and limits the number of panelists generally to one individual. Compared to these two methods, the finger span method overcomes these problems. The more rapid evaluation of the odor intensity allows to use a higher number of panelists and thus results can be treated by real statistical analysis. Good correlation was obtained between the finger span variables and the theoretic intensity calculated from bibliographic data. Ethylbutyrate used for training seems also particularly interesting for selection tests. This efficient method should be applied to other food aroma extracts and to "nose-space" experiments in progress in our laboratory.

Acknowledgments

This research was financed by TEPRAL, the research center from Kronenbourg (Danone, Strasbourg, France) and the Regional Council of Burgundy.

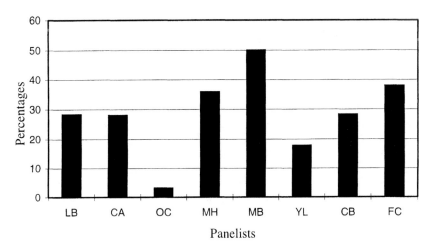

Figure 6. Comparison of the percentage of discriminating variables common with at least 4 panelists (LB, CA, OC, MH : trained panelists; MB,YL, CB, FC : untrained panelists).

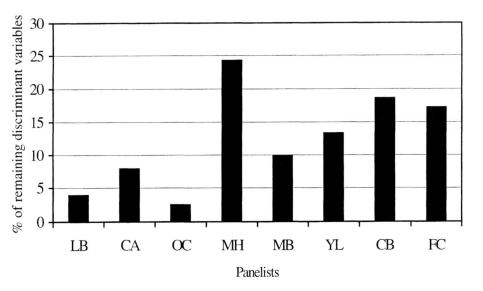

Figure 7. Decrease of the percentage of significant variables after elimination of nil values for each panelist (LB, CA, OC, MH : trained panelists; MB,YL, CB, FC : untrained panelists).

References

1. Acree, T.E. ; Barnard, J. ; Cunnungham, D.G. *Food Chem.* **1984**, *14*, 273-286.
2. Ullrich, F. ; Grosch, W. Z. *Lebensm. Unters. Forsch.* **1987**, *184*, 277-282.
3. Grosch, W. *Trends Food Sci. Technol.* **1993**, *4*, 68-73.
4. Mistry, B.S. ; Reineccius, T. ; Olson L. Techniques for Analysing Food Aroma Marsili, R., Ed. ; Dekker : New York, 1997 ; pp 265-292.
5. McDaniel, M.R. ; Miranda-Lopez, R. ; Watson, B.T. ; Michaels, N.J. ; Libbey, L.M *Proceedings of the 6th International Flavor Conference* ; Elsevier science Publishers Amsterdam, The Netherlands, 1989; pp 23-36.
6. da Silva, M.A.A.P. ; Lundahl, D.S. ; McDaniel, M.R. *Trends in Flavour Research* Maarse, H., Van der Heij, D.G., Eds ; Elsevier Science : Amsterdam, The Netherlands 1994, pp 191-209.
7. Etiévant, P.X. ; Callement, G. ; Langlois, D. ; Issanchou, S. ; Coquibus, N. *J. Agric. Food Chem.* **1999**, *47*, 1673-1980.
8. Cunningham, D.G. ; Acree, T.E. ; Barnard, J.; Butts, R.M. ; Braell, P.A. *Food Chem* **1986**, *19*, 137-14.
9. Stevens, S.S. *Psychophysics* ; Wiley : New York, 1975 ; pp 99-133.
10. Devos, M.; Rouault, J.; Laffort, P. **1998**, unpublished results.

Chapter 16

Aroma of Roasted Sesame Oil: Characterization by Direct Thermal Desorption–Gas Chromatography– Olfactometry and Sample Dilution Analysis

Keith R. Cadwallader[1] and Jeong Heo

Department of Food Science and Technology, Mississippi Agricultural Experiment Station, Mississippi State University, Mississippi State, MS 39762–5953
[1]Current address: Department of Food Science and Human Nutrition, University of Illinois at Urbana-Champaign, 202 Agricultural Bioprocess Laboratory, 11302 West Pennsylvania Avenue, Urbana, IL 61801

Aroma constituents of roasted sesame seed oils were isolated and identified by direct thermal desorption-gas chromatography-olfactometry (DTD-GCO) and DTD-GC-mass spectrometry (MS). Sample dilution analysis (SDA), in which decreased volumes of oil were sequentially analyzed by DTD-GCO, was used to indicate predominant odorants. Forty-nine odorants were detected by SDA with detection volumes (DVs) from 8 to 1000 nL. Those detected with DVs from 8 to 40 nL were 1-octen-3-one, 4,5-epoxy-(E)-2-decenal, 2-acetyl-3-methylpyrazine, 2-methoxyphenol, 2,3-diethyl-5-methylpyrazine, 3-methylbutanal, (E)-2-nonenal, 2-methoxy-4-vinylphenol, and an unidentified compound (*plastic aroma note*). Results of quantitative analysis were in general agreement with those of SDA.

In many parts of the world sesame (*Sesamum indicum* L.) is an important oil seed crop. In Europe and the United States, sesame in the form of the roasted seeds is popular on bread and bakery goods; while in Asia, roasted sesame seed oil is used as a condiment or seasoning agent in many dishes (*1*). During the manufacture of roasted sesame seed oil, sesame seeds are first roasted by various techniques to develop a characteristic roasted aroma. The conventional method involves cleaning of seeds followed by roasting, grinding, cooking, and pressing, but not refining (*2*). The color, composition, flavor, and oxidative stability of roasted sesame seed oil are all influenced by the conditions of the roasting process (*3-5*). The overall aroma note generated by roasting is often characterized as burnt, caramel, fatty, roasted, or meat-like (*6,7*).

The volatile components of roasted sesame seeds or the oil extracted from the roasted seeds has been extensively studied and more than 200 compounds have been identified (*4,6-19*). Only a few studies have been conducted on the sensory contribution of individual volatile compounds to the aroma of roasted sesame seeds (*7,16-18*) and oil (*16,19*).

The past two decades have experienced tremendous growth in the development and application of gas chromatography-olfactometry (GCO) in aroma research (*20-23*). Numerous GCO techniques are currently in use, with the most popular being the so-called 'dilution analysis' methods, which include aroma extract dilution analysis or AEDA (*22*) and CharmAnalysis[TM] (*23*). In both of these techniques a dilution series of an aroma extract is evaluated by GCO. For example, in AEDA the highest dilution at which a compound is smelled is defined as its flavor dilution factor, which is proportional to the odor activity value of the compound. A major limitation of such dilution techniques is that they do not account for highly volatile compounds that are lost during solvent extraction and work up procedures. Headspace-GCO techniques, such as GCO of decreasing headspace volumes (GCO-H) (*24*) and of dynamic headspace volumes (GCO-DHS) (*25*) have been used to augment data from AEDA to provide a more complete evaluation of food aroma composition.

Direct thermal desorption (DTD) combined with cryogenic trapping and GCO allows for the evaluation of essentially the whole range of volatiles, from those of very high volatility to the semivolatiles. It is possible to conduct 'dilution analysis' using this approach, in which stepwise dilutions (or decreased amounts or volumes) of a dry or semidry sample (e.g. an edible oil or dried spice) are analyzed by DTD-GCO. Results of DTD-GCO-sample dilution analysis (SDA) allow for ranking of odorants based on lowest volume (or mass) of sample required for their detection by DTD-GCO. Potential advantages of DTD include ease of sample preparation, need for only a small quantity of sample, and high sample throughput (*26*). DTD-GC is limited to analysis of low moisture foods and has been successfully employed in the analysis of edible oils and fats (*27-29*). In many cases, the results of DTD-GC compared favorably with sensory flavor scores of taste panels on the same products (*30,31*). DTD has some limitations in the analysis of food lipids, such as breakdown of thermally labile hydroperoxides (*26,29,32*). Despite this, DTD might be suitable for the analysis of roasted sesame seed oil, where the primary odorants are mostly *Maillard* reaction and *Strecker* degradation products and to a lesser extent lipid-derived compounds (*4,6-9,11-14, 16-19*).

The aim of this study was to evaluate the use of DTD-GCO and SDA for the determination of predominant odorants in roasted sesame seed oils. Volatile components of two Korean (commercially available) roasted sesame oils were isolated by DTD using a modified external closed loop inlet device and analyzed by GCO and SDA. Concentrations of selected odorants were determined by DTD-GC-MS to allow for comparison of their DVs with the calculated odor activity values.

Materials & Methods

Materials. Two commercial samples of roasted sesame seed oil were obtained from Haepyo Co., Ltd. (HC; Seoul, Korea) and Jeil Jedang Co., Ltd. (JJ; Inchon, Korea). An unroasted sesame seed oil sample was from Hain Food Group, Inc (Uniondale, NY). Samples were stored frozen (-20°C) until analyzed.

Reference compounds listed in Tables 1 and 2 were obtained from commercial sources. Compound no. 37 was synthesized using a published procedure (7). Distilled-deodorized water was prepared by boiling distilled water in a 4-L Erlenmeyer flask until volume was reduced by one-third.

Preparation of Deodorized Unroasted Sesame Seed Oil (USO). Fifty grams of silica gel (preconditioned at 160°C for 2h) was mixed with 2.5 mL of distilled-deodorized water in a sealed jar, shaken for 10 min and stored at room temperature for an overnight period. The silica gel was transferred to wine-glass funnel (6.2 cm length x 3.5 cm i.d.) and 100 g of USO passed through the silica gel bed. Deodorized oil (40 mL) was transferred to a 50-ml vial equipped with a PTFE cap, purged with nitrogen gas for 5 min, and stored at -20°C until analysis.

Direct Thermal Desorption-Gas Chromatography-Olfactometry (DTD-GCO) and Sample Dilution Analysis (SDA). A modified external closed inlet device (ECID; Scientific Instruments Services, Inc., River Ridge, LA) (Figure 1) was installed on an HP 5890 gas chromatograph (GC)(Hewlett-Packard Co., Palo Alto, CA). Column effluent was split 1:1 between a flame ionization detector (FID) and sniffing port by using deactivated fused silica transfer lines (0.25 mm i.d. x 1 m length). Separations were performed on either DB-5MS and DB-WAX columns (30 m length x 0.53 mm i.d. x 1.5 m (or 1 μm for DB-WAX) film thickness; J & W Scientific, Folson, CA). A silane-treated glass liner (5.8 cm length x 0.9 cm i.d.) was packed with volatile-free silane-treated glass wool, loose enough to permit diffusion of oil throughout the packing, yet tight enough to prevent seepage of the sample from the liner into the GC column. Clearance of about 2 cm was allowed at the bottom and top of liner. The liner was placed into the inlet device and the unit sealed with a septum. Carrier gas (helium) was directed (by positioning a six-port rotary valve) through the liner at a rate of 5 mL/min and then into the GC column. Prior to injection of a sample, a 15-cm section of the GC column was cooled in liquid nitrogen to cryofocus the volatiles. An oil sample (5, 1 or 0.2 μL of neat oil or a 0.2 μL aliquot of a 1:5 or 1:25 serial dilution (e.g. 1:5 v/v, one part oil + four parts of dichoromethane) was injected using a series 7101 syringe (Hamilton Co., Reno, NV) onto the top of the glass wool plug of the liner (27) maintained at ambient temperature (25-30°C). Immediately after injection, the inlet device was heated at a rate of 35°C/min and held at 150°C for 10 min, then the six-port rotary value was repositioned to allow for the helium to by-pass the inlet device, while still maintaining flow to the GC column. As quickly as possible, the GC column section was removed from the liquid nitrogen and the GC oven was rapidly heated. The run

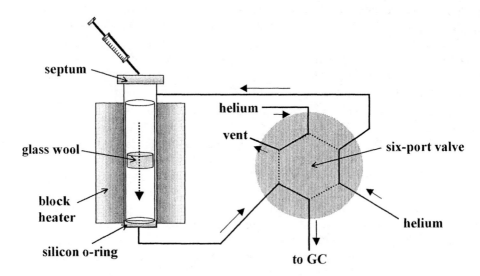

Figure 1. Modified external closed loop inlet device used for direct thermal desorption.

(GC) was started when the oven temperature reached 40°C. Oven temperature was programmed from 40° C to 225° C at a ramp rate of 6°C/min with initial and final hold times of 5 and 30 min, respectively. FID and sniffing port transfer lines were held at 250° C. Sniff port was supplied with humidified air at 30 mL/min.

The above inlet device temperature settings were determined experimentally by analyzing a 5 µL roasted sesame oil sample spiked with 1-tetradecanol (10 µg per g of oil). Desorption temperature of 150°C was chosen based on the literature, as this temperature is between that used previously for corn oils (120°C) and soybean and blended oils (170°C) (*31*). The final hold time was decided upon based on the recovery of the internal standard from the oil after hold times of 5, 10, and 20 min. A hold time of 10 min was found to be optimal, as this gave the best overall recovery in the shortest period of time (data not shown).

Each sample dilution was evaluated by two trained panelists who were instructed to record odor intensity (scale from 0 = no odor, to 7 = very strong), retention time, and odor properties of each odorant perceived during GCO. Results are expressed as detection volumes (DVs) on DB-5MS column. A DV is the lowest volume of sample required for detection of an odorant by DTD-GCO.

Direct Thermal Desorption-Gas Chromatography-Mass Spectrometry (DTD-GC-MS). A Tekmar[TM] 3000 Purge and Trap Concentrator/Cryofucusing Module (Tekmar Co., Cincinnati, OH) coupled to an HP5890 Series II GC/HP5972 mass selective detector (Hewlett-Packard Co.) was used for DTD-GC-MS. The Tekmar unit was configured to operate in the thermal desorption mode. A 30-cm length of 0.3-cm stainless steel glass-lined tubing (sample tube) was loosely packed with volatile-free silane-treated glass wool to permit diffusion of oil throughout the packing. Clearance of about 12.3 cm was allowed at the bottom of the liner and 2.5 cm at the top. The sample tube and glass wool packing were cleaned by baking at 225°C under a constant flow (40 mL/min) of helium.

For analysis, a 5 µL aliquot of an oil sample spiked with an internal standard was injected onto the top of the glass wool packing using a series 701 (Hamilton Co.) syringe. The spiked oil sample was prepared by adding 5 µL (4.07 ng) of an internal standard solution (2-methyl-3-heptanone in methanol) to 5 g of oil, followed by 5 min of vigorous shaking and settling (1 h) prior to analysis. The sampling tube was connected to the trap heater of the Tekmar unit. The volatiles were purged (with helium) from the oil sample at 150°C for 10 min and then subsequently cryofocused (-150°C) onto a 15-cm section of 0.53 mm i.d. deactivated fused silica capillary column. Transfer lines and valves were maintained at a temperature of 175°C. Helium flows during thermal desorption of sample tube (20 mL/min) and cryofocusing trap (1 ml/min) were controlled by the split/splitless electronic pressure control pneumatics of the GC as previously described (*35*). Cryofocused volatiles were thermally desorbed (180°C for 1 min) directly into the analytical GC column. Between each analysis, the DTD system was thoroughly baked out (225°C for 10 min) after installation of a clean sample tube.

GC separations were performed on DB-5MS and DB-WAX columns (60 m length x 0.25 mm i.d. x 0.25 μm film thickness; J&W Scientific). GC oven temperature was programmed from 40°C to 200°C at a rate of 3°C/min, with initial and final hold times of 5 and 60 min, respectively. MSD conditions were as follows: capillary direct interface temperature, 280°C; ionization energy, 70 eV; mass range, 33-350 a.m.u.; electron multiplier (EM) voltage, 200 V above autotune; scan rate, 2.2 scans/s. Compound identifications were based on comparison of retention indices (RI), mass spectra, and odor properties of unknowns with those of standard reference compounds analyzed under identical experimental conditions. Tentative identifications were based on matching RI values and odor properties of unknowns with those of reference compounds.

Quantitation of Aroma Compounds. Concentration of selected positively identified aroma compounds were determined using calibration curves of amount ratios (compound/internal standard (I.S.)) versus peak area ratios (compound/internal standard (I.S)). Standard solutions were prepared by spiking deodorized USO with the internal standard plus the standard reference compounds at three concentration levels and the spiked samples were analyzed by DTD-GC-MS as previously described.

Results and Discussion

The two roasted sesame seed oils used in this study were selected from a group of 10 commercial products on the basis of sensory evaluation (overall liking). Products originated from various geographic locations, including China, Japan, Korea, Singapore, England, and the United States. The majority of the panel members (composed of 22 individuals of five nationalities) noted that they preferred the aroma of the two Korean products (HC and JJ) because of their more intense and desirable roasted and nutty notes. Furthermore, these products were notably darker in color than the other samples, suggesting that they probably underwent a more extreme roasting process.

Volatile Profiles

Typical total ion chromatograms of the two commercial roasted sesame seed oils (HC and JJ) are shown in Figure 2. The analysis was performed by DTD-GC-MS on a 5 uL aliquot of oil. The chromatograms demonstrate a potential advantage of DTD for isolation of the wide range of volatiles, from the highly volatile acetaldehyde to the semivolatile compound 2-methoxy-4-vinylphenol, in roasted sesame seed oil. Excellent chromatographic resolution for most of the volatile constituents could be achieved on a DB-WAX column; however, the use of a DB-5MS column gave better results for some minor components, especially thiol derivatives (e.g. 2-furfurylthiol).

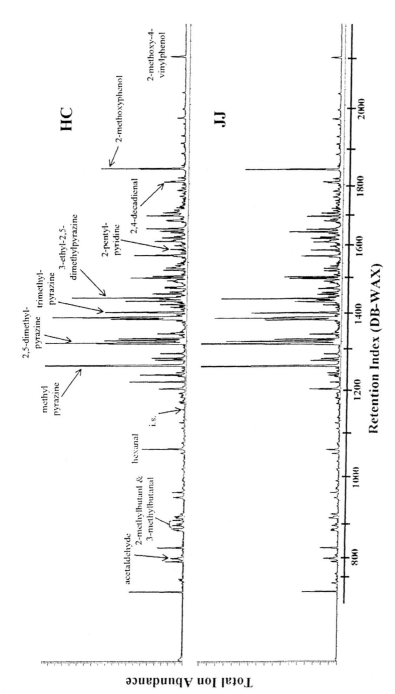

Figure 2. Total ion chromatograms of commercial roasted sesame seed oils.

Alkylpyrazines were in highest abundance among the identified volatile constituents. These *Maillard* reaction products have been reported to impart nutty, roasted aromas to foods (*34*). The formation and occurrence of pyrazines has been extensively studied in numerous heated food products and model systems (*34, 35*). However, with only a few exceptions, alkylpyrazines are not considered to be key odorants in foods because of their high odor detection thresholds. Therefore, their high abundance in roasted sesame seed oil does not necessarily indicate their predominance in the aroma profile. In addition to the alkylpyrazines several other *Maillard* reaction products were found in relatively low abundance, including numerous pyrroles, pyridines and thiazoles, as well as S*trecker* aldehydes. 2-Methoxyphenol (guaiacol) and 2-methoxy-4-vinylphenol, were found in relatively high abundance. Ketones, aldehydes, alcohols, alkanes and other compounds originating from oxidation and thermal degradation of lipids were in moderately high abundance. Overall, the above findings were in agreement with previous reports on the volatile composition of roasted sesame oil (*4,14,19*). It is evident from the results of the present study, as well as from previous reports (*11,15,16,17*) that the volatile products formed during the roasting of sesame seeds originate from several pathways, such as the Maillard reaction, thermal degradation of phenolic (e.g. vanillic acid) precursors, and oxidative and thermal degradation of lipids.

Sample Dilution Analysis

Results of DTD-GCO and SDA reveal a total of 49 odorants detected in the DV range of 8 to 1000 nL in at least one of the two commercial oil samples (Table 1). Nine compounds (nos. 6, 20, 26, 30, 33, 34, 44, 47 and 48) were detected in both oils with low DVs from to 8 to 40 nL Because of their low DVs, these compounds should have a greater impact on the aroma of the roasted sesame seed oils. 1-Octen-3-one (no. 20; *mushroom*) and a compound tentatively identified as 4,5-epoxy-(*E*)-2-decenal (no. 47; *metallic, fatty*) were detected in both oil samples at a DV of 8 nL. 2-Acetyl-3-methylpyrazine (no. 30; *popcorn, coffee*), 2-methoxyphenol (no. 31; *smoky*) and 2,3-diethyl-5-methylpyrazine (no. 34; *nutty, potato*) were found at a DV of 8 nL in sample JJ and at a DV of 40 nL in sample HC . An unknown compound (no. 39; *coffee*) was found with a DV of 8 nL in JJ, but was found at a much higher DV of 5000 nL in HC. 3-Methylbutanal (no. 6; *malty, chocolate*), (E)-2-nonenal (no. 33; *stale, hay*), 2-methoxy-4-vinylphenol (no. 44, *smoky, cloves*), β-damascenone (no. 48; *tea, applesauce*), and an unknown compound (no. 26; *plastic*) were detected at a DV of 40 nL in both oil samples. Compounds found at moderate intensities (DV=40 to 200 nL) in both oil samples included acetaldehyde (no. 1; *pungent, yogurt*), methylpropanal (no. 4; *malty, chocolate*), heptanal (no. 14; *sweet, ethanolic*), octanal (no. 22; *orange oil, sweet*), 2-acetyl-2-thiazoline (no. 32; *roasted, popcorn*), 2-pentylpyridine (no. 37; *fatty, metallic*), (*E*)-2-undecenal (no. 46; *waxy, cilantro*) and one compound tentatively identified as 2-methyl-(3-methyldithio)furan (no. 36, *vitamin, meaty*). In addition, two unknown compounds, no. 23 (*pungent, plastic*) and no. 40 (*stale, bitter*) were detected in this DV range. Hexanal (no. 9;

green, cut-grass) was found at a DV of 40 nL in JJ, but was at lower intensity in HC (DV = 1000 nL). The remaining 27 compounds (nos. 2, 3, 5, 7, 8, 10-13, 15-19, 21, 24, 25, 27-29, 35, 38, 41-43, 45, and 49) in Table 1 were detected at DVs ranging from 200 to 5000 nL.

The results of SDA demonstrate the involvement and importance of several reaction pathways (as mentioned earlier) in the development of the typical aroma constituents of roasted sesame seed oil. The origins of these compounds in roasted sesame oil have been previously discussed (*16*). The number, type and relative aroma intensities for most compounds in Table 1 are similar to what was reported earlier by Schieberle (*7, 16,17*) for roasted sesame seeds; however, some notable differences were observed. For example, 2-furfurylthiol (no. 17), 2,5-dimethyl-4-hydroxy-3(*2H*)-furanone (no. 28), and 2-acetyl-1-pyrroline were previously indicated as intense odorants in roasted white and black sesame seeds (*7,16,17*), while in the present study these compounds were detected at relatively low intensities in both oil samples. (The results for 2-acetyl-1-pyrroline are not shown in Table 1. This compound was detected in both oil samples at a DV of 5000 nL.) These compounds may not have been efficiently isolated by DTD or possibly they underwent thermal decomposition. On the other hand, it is possible that the commercial roasted sesame oils contained lower levels of these polar compounds because of their poor extractability, solubility, or stability in the nonpolar oils. Two lipid-derived compounds, 1-octen-3-one (no. 20) and 4,5-epoxy-(*E*)-2-decenal (no. 47), were found at high odor intensities in the two oil samples. It is possible that these compounds increased during DTD as result of thermal decomposition of hydroperoxides, or possibly these constituents are naturally present at higher levels in the oil than would be expected for the intact seeds. Finally, four highly volatile odorants (nos. 1-4) were detected in the present study which were not previously reported as predominant odorants in roasted sesame seeds and oils.

Concentrations and Odor Activity Values

Concentrations and odor activity values (OAVs) for selected volatile constituents of the two roasted sesame seed oil samples are given in Table II. These findings are in general agreement with previous reports on the volatile composition of roasted sesame seed oil (*14,19*). Alkylpyrazines were the major volatile components of the oils, with 2-methylpyrazine in highest abundance followed closely by 2,5-dimethylpyrazine, 2,6-dimethylpyrazine and 3-Ethyl-2,5-dimethylpyrazine. 2-Methoxyphenol was the second most abundant volatile constituent of the oils. *Strecker* aldehydes (nos. 1,4,6,7) were detected at moderately high concentrations, while the lipid-derived compounds pentanal, hexanal and 1-octen-3-one were found at comparatively low levels in both oils. 2-Furfurylthiol was not included in Table II because it was found at only trace levels.

The OAVs in Table II were calculated on the basis of odor detection thresholds determined in oil. The odor detection thresholds for most of alkylpyrazines are

Table I. Predominant Odorants in Roasted Sesame Seed Oils by Direct
Thermal Desorption-Gas Chromatography-Olfactometry and
Sample Dilution Analysis

No[a]	Compound name	Ret. Index[b] DB5MS	DBWAX	Odor property	Det.Vol.(nL)[c] HC	JJ
1	Acetaldehyde	<500	618	pungent, yogurt	200	200
2	Methanethiol	<500	655	sulfurous, rotten	5000	1000
3	Unknown	<500	- -	amine, fishy	1000	200
4	Methylpropanal	560	813	malty, chocolate	200	200
5	2,3-Butanedione	613	986	cream cheese	1000	5000
6	3-Methylbutanal	654	917	malty, chocolate	40	40
7	2-Methylbutanal	663	906	malty, musty	1000	200
8	3-Methylthiophene	779	1110	pungent, plastic	1000	5000
9	Hexanal	802	1071	green, cut-grass	1000	40
10	3-Methylbutyric acid	835	1662	dried fruit	5000	1000
11	2-Methylbutyric acid	837	1662	malty, musty	nd	1000
12	Unknown	861	- -	garlic salt	200	5000
13	Unknown	871	- -	pungent, styrene	1000	200
14	Heptanal	890	1178	sweet, ethanolic	40	200
15	Unknown	902	- -	popcorn	1000	200
16	3-(Methylthio)propanal	911	1461	cooked potato	5000	1000
17	2-Furfurylthiol[d]	922	1436	roasted, coffee	5000	1000
18	[4-methyl-3-thiazoline][e]	956	- -	garlic	5000	200
19	Unknown	959	- -	earthy, peanut	5000	200
20	1-Octen-3-one	978	1291	mushroom	8	8
21	2-Ethyl-(5or6)-methyl-pyrazine	995	1380	ethanolic, sweet	5000	200
22	Octanal	1005	1289	orange oil, sweet	40	200
23	Unknown	1017	- -	pungent, plastic	200	200
24	2-Acetylthiazole	1020	1653	roasted, popcorn	1000	200
25	2-Acetylpyrazine	1029	1634	roasted, popcorn	1000	200
26	Unknown	1044	- -	plastic	40	40
27	Phenylacetaldehyde	1050	1656	floral, hay	5000	1000
28	2,5-Dimethyl-4-hydroxy-3(2H)furanone	1070	2070	burnt sugar	1000	200
29	2-Ethyl-3,5-dimethyl-pyrazine	1084	1467	earthy, nutty	5000	1000

Table I. Continued

No[a]	Compound name	Ret. Index[b] DB5MS	DBWAX	Odor property	Det.Vol.(nL)[c] HC	JJ
30	2-Acetyl-3-methyl-pyrazine	1089	1629	popcorn, coffee	40	8
31	2-Methoxyphenol	1097	1863	smoky	40	8
32	2-Acetyl-2-thiazoline	1109	1773	roasted, popcorn	40	200
33	(E)-2-Nonenal	1152	1510	stale, hay	40	40
34	2,3-Diethyl-5-methyl-pyrazine	1165	1490	nutty, potato	40	8
35	unknown	1181	- -	stew, meaty	1000	1000
36	[2-Methy-(3-methyldi-thio-furane][e]	1185	- -	vitamin, meaty	200	40
37	2-Pentylpyridine	1203	1527	fatty, metallic	200	40
38	(E,E)-2,4-Nonadienal	1224	1702	fatty, fried	1000	40
39	unknown	1230	- -	coffee	5000	8
40	unknown	1255	- -	stale, bitter	40	200
41	(E)-2-Decenal	1268	1641	green, cilantro	1000	200
42	unknown	1275	- -	stale, fatty	1000	200
43	unknown	1285	- -	fatty, cinnamon	5000	1000
44	2-Methoxy-4-vinyl-phenol	1312	2193	smoky, cloves	40	40
45	(E,E)-2,4-Decadienal	1324	1803	fatty, fried	5000	1000
46	(E)-2-Undecenal	1351	1744	waxy, cilantro	200	200
47	[4,5-Epoxy-(E)-2-decenal][e]	1387	- -	metallic, fatty	8	8
48	β-Damascenone[d]	1396	1821	tea, applesauce	40	40
49	3-Methylindole	1427	2401	pungent, fecal	1000	200

[e]Numbers correspond to those in Tables 2 and Figure 2. [b]Retention index determined from GCO data. [c]Detection volume, lowest volume of sample required for detection of odorant by DTD-GCO on DB5ms column. [d]Mass spectrum unavailable, compound tentatively identified based on RI values and odor property. [e]Tentatively identified based on published RI value and odor property (7).

Table II. Concentrations and Odor Activity Values of Selected Volatile Components of Roasted Sesame Seed Oils

No.[a]	Compound name	Concentration (ng/g)[b]		Odor Detection Threshold (ppb)[c]	Odor Activity Value[d]	
		HC	JJ		HC	JJ
1	Acetaldehyde	2600 (±11%)	2900 (±28%)	0.22[e]	11800	13200
4	Methylpropanal	678 (±20%)	914 (±11%)	43[f]	16	21
6	3-Methylbutanal	4200 (±22%)	2780 (±64%)	13[f]	323	214
7	2-Methylbutanal	3860 (±22%)	2180 (±36%)	140[f]	28	16
	Pentanal	318 (±24%)	170 (±49%)	--	--	--
9	Hexanal	408 (±12%)	355 (±45%)	120[g]	3	3
20	Methylpyrazine	32700 0(±20%)	74600 (4±%)	--	--	--
	1-Octen-3-one	782 (±47%)	553 (±50%)	10[h]	78	55
	2,5-Dimethylpyrazine	6870 (±47%)	16000 (±25%)	2600[g]	3	6
	2,6-Dimethylpyrazine	3910 (±44%)	11400 (±48)	--	--	--
	Ethylpyrazine	1920 (±7)	4650 (±38)	--	--	--
	2,3-Dimethylpyrazine	1380 (±56%)	2630 (±24%)	--	--	--
21a	2-Ethyl-6-methyl-pyrazine	2870 (±25%)	7190 (±47%)	--	--	--
21b	2-Ethyl-5-methyl-pyrazine	1930 (±9%)	4640 (±66%)	320[g]	6	14
	Trimethylpyrazine	1110 (±52%)	2120 (±38%)	--	--	--

3-Ethyl-2,5-dimethylpyrazine	3207 (±50%)	6870 (±38%)	24[g]	134	286
29 2-Ethyl-3,5-dimethylpyrazine	611 (±64%)	1310 (±41%)	2.2[i]	278	595
34 2,3-Diethyl-5-methylpyrazine	296 (±49%)	472 (±44%)	- -	- -	- -
3,5-Dimethyl-2-methylpyrazine	482 (±34%)	1180 (±56%)	- -	- -	- -
37 2-Pentylpyridine	582 (±67%)	360 (±42%)	5[i]	116	72
24 2-Acetylthiazole	422 (±34%)	996 (±10%)	- -	- -	- -
25 2-Acetylpyrazine	560 (±33%)	1280 (55±%)	10[i]	56	128
30 2-Acetyl-3-methylpyrazine	242 (±33%)	640 (±19%)	- -	- -	- -
32 2-Acetyl-2-thiazoline	145 (±4%)	215 (±13%)	- -	- -	- -
45 (E,E)-2,4-Decadienal	324 (±70%)	332 (±56%)	135[g]	2	2
31 2-Methoxyphenol	11700 (±65%)	15000 (±7)	19[i]	616	789
44 2-Methoxy-4-vinylphenol	2260 (±6%)	4420 (±41%)	50[i]	45	88

[a]Numbers correspond to those in Table 1. [b]Average concentration (± percent relative standard deviation) of duplicate determinations. [c]Orthonasal detection threshold in vegetable oil. [d]Odor activity value = average concentration divided by odor detection threshold. [e]ref 36. [f]ref 37. [g]ref 38. [h]ref 39. [i]ref 7.

relatively high and, therefore, these compounds generally have low OAVs, indicating that they may play only minor roles in the aroma of roasted sesame seed oil. Based on OAVs, the most intense odorants in the roasted sesame seed oils include acetaldehyde, 2-methoxyphenol, 2-ethyl-3,5-dimethylpyrazine, 3-methylbutanal, 2-pentylpyridine, 1-octen-3-one, and 2-acetylpyrazine. However, it is difficult to make conclusions beyond these compounds for lack of quantitative information and unavailability of odor threshold values (in oil) for many of the other odorants detected by SDA.

Advantages and Disadvantages of DTD-GCO and SDA

SDA offers some potential advantages as a 'dilution analysis' technique for the determination of predominant odorants in dry or semidry foods, such as edible oils. Perhaps the biggest advantage of SDA is that compounds of widely varying volatilities can be simultaneously sampled and evaluated. SDA is a relatively simple and rapid method that requires only a small sample size and minimal or no sample preparation. No solvent extraction, distillation, or cleanup step is required Quantitative analysis of highly volatile constituents by DTD-GC-MS is more easily accomplished because there is no interference from solvent or solvent impurity peaks (*40*). Despite its several advantages, there are some definite disadvantages to the SDA approach. Most notable is the potential for loss of thermally labile constituents during DTD. Furthermore, with edible oils there is a greater chance for decomposition of hydroperoxides during DTD, especially if high desorption temperatures are used. In DTD, a tradeoff must be made between desorption temperature, heating rate, and sampling time (*26,40*). Of particular importance is the desorption temperature, which must be high enough to allow for efficient mass transfer of the volatile constituents while being maintained low enough to minimize sample decomposition. It is therefore recommended that special consideration be taken when evaluating results of SDA, since the method evaluates both stable odorants as well as some thermal decomposition products.

Literature Cited

1. Namiki, M. Sesame. *Food Reviews International* **1995**, *11*, 281-329.
2. Fukuda, Y.; Namika, M. *Nippon Shokuhin Kogyo Gakkaish* **1988**, *35*, 552-562.
3. Yen, G.C. *J. Sci. Food Agric.* **1990**, *50*, 563-570.
4. Shimoda, M.; Nakada, Y.; Nakashima, M.; Osajima, Y. *J. Agric. Food Chem.* **1997**, *45*, 3193-3196.
5. Yoshida, H.; Takagi, S. *J. Sci. Food Agic.* **1997**, *75*, 19-26
6. Manley, C.H.; Vallon, P.P.; Ericson, R.E. J. Food Sci. **1974**, *39*, 73-76.
7. Schieberle, P. *Food Chem.* **1995**, *55*, 145-152.
8. Kinoshita, S.; Yamanishi, T. *Nippon Nôgeikagaku Kaishi* **1973**, *47*, 737-739.

9. Soliman, M.M.; Kinoshita, S.; Yamanishi, T. *Agric. Biol. Chem.* **1975**, *39*, 973-977.
10. Soliman, M.M.; El-Sawy, A.A.; Fadel, H.M.; Osmon, F. *J. Agric. Food Chem.* **1985**, *33*, 523-528.
11. Soliman, M.M.; El-Sawy, A.A.; Fadel, H.M.; Osman, F. *Acta Alimentaria* **1986**, *15*, 251-263.
12. El-Sawy, A.A.; Soliman, M.M.; Fadel, H.M. *Grasas y Aceites* **1988**, *39*, 160-162.
13. El-Sawy, A.A.; F. Osman, F.; Fadel, H.M. *Seinfen Öle Fette Wachse* **1989**, *115*, 667-679.
14. Nakamura, S.; Nishimura, O.; Masuda, H.; Mihara, S. *Agric. Biol. Chem.* **1989**, *53*, 1891-1899.
15. Takei, Y. *Nippon Kasei Gakkaishi* **1989**, *40*, 23-34.
16. Schieberle, P. In *Progress in Flavor Precursor Studies*; Schreier, P.; Winterhalter, P., Eds.; Allured: Carol Stream, IL, 1993; pp. 343-360.
17. Schieberle, P. In *Olfaction and Taste XI*; Kurihara, K.; Suzuki, N.; Ogawa, H., Eds.; Springer-Verlag: New York, 1994; pp. 263-267.
18. Schieberle, P..; Güntert, M.; Sommer, H.; Werkhoff, P. *Food Chem.* **1996**, *56*, 369-372.
19. Shimoda, M.; Shiratsuchi, H.; Nakada, Y.; Wu, Y.; Osajima, Y. *J. Agric. Food Chem.* **1996**, *44*, 3909-3912.
20. Acree, T. In *Flavor Measurement.* Ho, C.-T.; Manley, C.H., Eds.; Marcel Dekker, Inc.: New York, 1993; pp. 77-94.
21. Blank, I. In *Techniques for Analyzing Food Aroma*; Marsili, R., Ed.; Marcel Dekker, Inc.: New York, 1997; pp. 293-329.
22. Grosch, W. *Trends in Food Science & Technology* **1993**, *4*, 68-73.
23. Acree, T. *Analytical Chemistry News & Features*, **1997**, *March 1*, pp. 170A-175A.
24. Milo, C.; Grosch, W. *J. Agric. Food Chem.* **1995**, *43*, 459-462.
25. Cadwallader, K.R.; Baek, H.H. In *Proceedings of the Ninth International Flavor Conference, George Charalambous Memorial Symposium*; Mussinan, C.; Contis, E.; Ho, C.-T.; Parliment, T.; Spanier, A.; Shahidi, F., Eds.; Elsevier Science B.V.: Amsterdam, 1997; pp. 271-279.
26. Hartman, T.G.; Lech, J.; Karmas, K.; Salinas, J.; Rosen, R.T.; Ho, C.-T. In *Flavor Measurement.* Ho, C.-T.; Manley, C.H., Eds.; Marcel Dekker, Inc.: New York, 1993; pp. 37-60.
27. Dupuy, H.P.; Flick, G.J.; Bailey, M.E.; St. Angelo, A.J.; Legendre, M.G.; Sumrell, G. *J. Am. Oil Chem. Soc.* **1985**, *62*, 1690-1693.
28. Suzuki, J.; Bailey, M.E. *J. Agric. Food Chem.* **1985**, 33, 343-347.
29. Snyder, J.M.; Frankel, E.N.; Selke, E.; Warner, K. *J. Am. Oil Chem. Soc.* **1988**, *65*, 1617-1620.
30. Dupuy, H.P.; Rayner, E.T.; Wadsworth, J.I. *J. Am. Oil. Chem. Soc.* **1976**, *53*, 628-631.
31. Dupuy, H.P.; Rayner, E.T.; Wadsworth, J.I.; Legendre, M.G. *J. Am. Oil Chem. Soc.* **1977**, *54*, 445-449.

32. Lee, I.; Fatemi, S.H.; Hammond, E.G.; White, P.J. *J. Am. Oil Chem. Soc.* **1995**, *72*, 539-546.
33. Cadwallader, K.R.; Howard, C.L. In *Flavor Analysis: Developments in Isolation and Characterization.* Mussinan, C.J.; Morello, M.J., Eds.; ACS Symposium Series 705; American Chemical Society: Washington, DC, 1998; pp. 343-358.
34. Maga, J.A.; Sizer, C.E. *J. Agric. Food Chem.* **1973**, *21*, 22-30.
35. Amrani-Hemaimi, M.; Cerny, C.; Fay, L.B. *J. Agric. Food Chem.* **1995**, *43*, 2818-2822..
36. Reiners, J.; Grosch, W. J. *J. Agric. Food Chem.* **1998**, *46*, 2754-2763.
37. Guadagni, D.G.; Buttery, R.G.; Turnbaugh, J.G. J. *Sci. Food Agric.* **1972**, *23*, 1435-1444.
38. Buttery, R.G.; Guadagni, D.G.; Ling, L.C. *J. Agric. Food Chem.* **1973**, *21*, 198-201.
39. Guth, H.; Grosch, W. *Lebensm.-Wiss. u.-Technol.* **1990**, *23*, 513-522.
40. Wampler, T.E. *LC·GC* **1998**, *16*, 812-821.

Author Index

Subject Index